DATE			

COMMUNICATION BY DESIGN

COMMUNICATION
BY
DESIGN

*The Politics of Information and
Communication Technologies*

Edited by

ROBIN MANSELL
and
ROGER SILVERSTONE

OXFORD UNIVERSITY PRESS
1996

Oxford University Press, Walton Street, Oxford OX2 6DP

Oxford New York

Athens Auckland Bangkok Bombay
Calcutta Cape Town Dar es Salaam Delhi
Florence Hong Kong Istanbul Karachi
Kuala Lumpur Madras Madrid Melbourne
Mexico City Nairobi Paris Singapore
Taipei Tokyo Toronto
and associated companies in
Berlin Ibadan

Oxford is a trade mark of Oxford University Press

Published in the United States
by Oxford University Press Inc., New York

British Library Cataloguing in Publication Data
Data available

Library of Congress Cataloging in Publication Data
Communication by design: the politics of communication technologies/
edited by Robin Mansell and Roger Silverstone.
Includes bibliographical references and index.
1. Telecommunication policy. 2. Telecommunication—Social
aspects. 3. Information technology—Social aspects. I. Mansell,
Robin. II. Silverstone, Roger.
HE7645.C654 1996 95–46910
384'.068—dc20
ISBN 0–19–828941–3

1 3 5 7 9 10 8 6 4 2

Typeset by Cambrian Typesetters, Frimley, Surrey
Printed in Great Britain
on acid-free paper by
Biddles Ltd., Guildford & King's Lynn

ACKNOWLEDGEMENTS

We are grateful to all the colleagues who have helped to stimulate our thinking throughout the eight-year Economic and Social Research Council (ESRC) Programme on Information and Communication Technologies (PICT). Without the sustained opportunity to engage in debate and discussion in disciplinary and multidisciplinary territories, our various perspectives would not have coalesced within the covers of this book. We are grateful also to the ESRC and the public and private sponsors of complementary research during the PICT programme. Many people in government and industry participated in our research and we thank them for their contributions.

Cynthia Little, secretary to the Science Policy Research Unit (SPRU), Centre for Information and Communication Technologies at the University of Sussex, contributed so substantially to the editing of chapters and to the processing of the manuscript that she really should be counted among the authors. Her diligence and encouragement to all of us have been exceptional and we cannot thank her enough. Extensive comments by W. Edward Steinmueller, Professor of the Economics of Technical Change, Maastricht Economic Research Institute on Innovation and Technology, The Netherlands; Professor Liora Salter, Director, Program in Law and the Determinants of Social Ordering, Canadian Institute for Advanced Research, Osgoode Hall Law School/Faculty of Environmental Studies, York University, Canada; and Dr Puay Tang, SPRU Research Fellow, have also been very much appreciated. Professor Michael Gibbons, SPRU Director, succeeded in provoking us into exploring theoretical perspectives beyond the immediate boundaries of our own academic institutions and we thank him for this. Professor Richard Lipsey, Program in Economic Growth: Science and Technology and Institutional Change in a Global Economy, Canadian Institute for Advanced Research, Simon Fraser University, Canada, offered Robin Mansell a working environment that was very conducive to the initial drafting stages of this book.

We also thank Professor William Melody, the first director of PICT, for his persistence in enabling the ESRC to launch a multidisciplinary research programme into the social and economic implications of information and communication technologies. Multidisciplinary inquiry is always challenging and we have found it so. All the contributors wondered at various times whether it might not be easier to keep our research safely bounded by more familiar disciplinary categories. In the end we chose to risk an adventure into more eclectic and uncharted territory. We learned a great deal in the process and hope we have succeeded in conveying our insights to our readers.

We dedicate this book to Professor Christopher Freeman, founding Director of SPRU, and initiator of SPRU's contribution to PICT. His conviction that advanced information and communication technologies can be made to contribute to a better world has encouraged us in our inquiry into how this might come to be. Any errors and omissions are, of course, our own responsibility.

Robin Mansell

University of Sussex at Brighton   Roger Silverstone

CONTENTS

LIST OF FIGURES AND TABLES

LIST OF CONTRIBUTORS

Dr Leslie Haddon is Senior Research Fellow, Graduate Research Centre in Culture and Communication, University of Sussex.

Dr Richard Hawkins is Research Fellow, Centre for Information and Communication Technologies, Science Policy Research Unit (SPRU), University of Sussex.

Dr Robin Mansell is Professor of Information and Communication Technology Policy, and Head of the Centre for Information and Communication Technologies at the Science Policy Research Unit (SPRU), University of Sussex, one of six centres funded by the United Kingdom's Economic and Social Research Council Programme on Information and Communication Technologies (PICT).

Paul Quintas is Senior Lecturer, School of Management, The Open University. Until June 1994 he was Senior Research Fellow at the Centre for Information and Communication Technologies, Science Policy Research Unit (SPRU), University of Sussex.

Dr Rohan Samarajiva is Associate Professor and Director of the Graduate Studies Programme in the Department of Communication, Ohio State University, Columbus, Ohio.

Dr Roger Silverstone is Professor of Media Studies and Director of the Graduate Research Centre in Culture and Communication, University of Sussex. Professor Silverstone is an Associate Fellow of the Science Policy Research Unit (SPRU).

ACRONYMS

4GL	Fourth Generation Language
ANSI	American National Standards Institute
ARPANET	Advanced Research Projects Agency Network (US DoD)
AT&T	American Telegraph & Telephone Company
BSI	British Standards Institution
CASE	Computer-Aided Software Engineering
CCSS7	Common Channel Signalling System No. 7
CD-i	Compact Disc Interactive
CD-ROM	Compact Disc-Read Only Memory
DEC	Digital Equipment Corporation
ECU	European Currency Unit
EDP	Electronic Data Processing
ENIAC	1946 computer developed at University of Pennsylvania
FCC	Federal Communications Commission, United States
GATT	General Agreement on Tariffs and Trade
GSM	Global System for MobileCommunications
i-Case	Integrated CASE
IBM	International Business Machines
INFO-2	AT&T service tariff
IPSE	Integrated Project Support Environment
ISDN	Integrated Services Digital Network
ISO	International Organization for Standardization
ITU	International Telecommunication Union
MCI	Microwave Communications Incorporated
MIT	Massachusetts Institute of Technology
NSFNET	National Science Foundation Network
OECD	Organisation for Economic Cooperation and Development
OSI	Open Systems Interconnection
PC	Personal Computer
R&D	Research and Development
SQL	Structured Query Language
SSADM	Structured Systems Analysis and Design Method

T1	Telecommunication standards committee, ANSI (Accredited Standards Committee I: Telecommunications)
TCP/IP	Transmission Control Protocol/Internet Protocol
TGI	Transaction-Generated Information
TRIPS	Trade-Related Intellectual Property Rights
TTC	Telecommunications Technology Committee (Japan)
TTGI	Telecommunication Transaction-Generated Information
UNCTAD	United Nations Conference on Trade and Development
UNESCO	United Nations Education, Scientific and Cultural Organization
UNIX	Portable computer operating system originally developed by Bell Laboratories
UPT	Universal Personal Telecommunications
VPN	Virtual Private Network
WARC	World Administrative Radio Conference
WATTC	World Administrative Telegraph and Telecommunication Conference
WIPO	World Intellectual Property Organization
WTO	World Trade Organization

Introduction

ROBIN MANSELL AND ROGER SILVERSTONE

> Once the inevitabilities are challenged, we begin gathering our
> resources for a journey of hope. If there are no easy answers there
> are still available and discoverable hard answers, and it is these that
> we can now learn to make and share.
>
> (Williams 1983: 268–9)

The information and communication technologies of the 1990s enable
the electronic production and consumption of increasingly vast quantities
of information. These technologies are becoming central to the production and use of information relevant to all aspects of business and
consumer, educational, and leisure activity. These changes are not
inconsequential. Their effects are likely to be both unpredictable and
contradictory. They raise profound concerns about the way advanced
information and communication technologies influence industry,
government policy, and our everyday lives. They raise profound
concerns about access and accessibility, and about their capacity to
contribute to a more equitable and democratic society. No longer
approachable through the uniform visions of singular disciplines, this
complex of technological and social interdependencies requires a
multidisciplinary perspective. Such a perspective, grounded in the
study of the politics, economics, and sociology of information and
communication technologies, is developed in the pages that follow.
Within it we hope to offer insights into the ways in which the legacies of
the history of technical and institutional innovation and the creative
capabilities of human agents structure the trajectories of the information
and communication technologies that are being developed and used in
the late twentieth century.

The relationships between the use of information and communication
technologies and the economic, political, and cultural significance of
communication and information have long been subjects of social
science investigation (Innis 1950). These studies have focused on those

The authors particularly want to thank Professor W. Edward Steinmueller, Maastricht
Economic Research Institute on Innovation and Technology, The Netherlands, for his
detailed and helpful comments on an earlier draft of this chapter. Our conceptual
approach and the structure it provides for subsequent chapters evolved as a result of a
series of meetings over two years among all the contributing authors. Their collective input
is gratefully acknowledged as are comments by Dr Puay Tang, Research Fellow, Science
Policy Research Unit.

technologies which have been seen, most often at the time as well as with the benefit of hindsight, as having potentially revolutionary significance. The printing press, the telegraph, radio broadcasting, the Whirlwind and ENIAC computers, and the first black-and-white broadcast television receiving sets have each in their turn been the subject of considerable critical attention (Rosenberg 1994*b*). Today this attention is being directed towards microprocessors and semiconductors, 'intelligent' networks, multimedia, and virtual reality applications, and urgently so.

Like their predecessors, advanced information and communication technologies have become intimate parts of our social and economic lives. They bring us information about the world around us and we engage with information and communication systems in an increasing proportion of our waking (and occasionally sleeping) lives. The generation which has grown up since the end of World War II was the first to experience global events through the increasingly insistent and immediate exposure to visual images in their living-rooms. It was the first to contemplate the feasibility of the remote design of aeronautical engines and other components of large technical systems by computer-aided technologies. This same generation also became aware that the new technological system made possible by tele-communication services, computers, and software developments might one day just as possibly announce the last few minutes of their lives as a result of the prospect of nuclear incineration or the degradation of the environment as proclaim that scientific research had yielded a new life-saving vaccine.

Advances in information and communication technologies have therefore been and still are associated with radical, and even revolution-ary, socio-economic change. These technologies have become engrained more pervasively, but still unevenly, in the fabric of our business and consumer cultures. They are affecting all aspects of our public and private lives. The innovation process involves a complex array of interdependencies between producers and users. We need research adequate to this complexity if we are to understand technology producer and user practices and to inform the policy formation and implementa-tion process.

Information and communication technologies are embedded in networks and services and these affect whether a growing proportion of the global stock of knowledge can be accessed and used as well as the local and global accumulation and flows of public and private know-ledge. Electronic networks embody complex hardware and software linked together by a vast array of technical protocols. Microelectronics, optical fibre, and radio-based technologies facilitate the processing and transmission of analogue and digital signals. These signals, in their various combinations and representations, have become an increasingly predominant 'currency' in today's economic, political, and social relationships.

Some analysts speculate on the emergence of 'intelprises' based on hypernetworks and co-operative information and knowledge sharing (Harasim 1993). Others expect the emergence of a surveillance society controlled by technically proficient élites who design and operate networks with global reach (Gandy 1993; Samarajiva 1996 forthcoming). Although the diffusion of advanced information and communication technologies creates the possibility that all people will be included within their ambit and that this will bring benefits, there are those who suggest that structural privileges will continue to exclude people from productive participation in electronically mediated environments (Scharpf 1993: 155).

When the veil of technological inevitability is challenged, as Raymond Williams suggests, we begin to see that information and communication technologies are being employed by producers and users in ways that depend on and alter highly culturally specific understandings about how communication relationships and the production and exchange of information are integrated within social, political, and economic life. Simplistic utopian or dystopian visions of the future provide us neither with an understanding of how these changes come about nor with an understanding of the longer-term implications. We believe, and argue here, that analysis of the realistic scope for choice available to technology producers and users and of the degrees of freedom available to them to mould the vast technological system encompassed by past, present, and future generations of information and communication technologies is needed in order to move beyond these visions.

Whether in the sphere of electronic commerce, public discourse, or entertainment, the generation, distribution, and use of information and knowledge occur more and more frequently in electronic environments. Multimedia and information highways have focused attention on the potential benefits perhaps rather more than on the problems associated with the diffusion of advanced information and communication technologies. Our central concern in this book is with the analysis of the causes and consequences of alternative technical trajectories for the deployment of what Freeman and Soete (1994: 39) have called 'the biggest technological juggernaut that ever rolled'. Our aim is to illustrate how the innovation process gives rise to the diverse and dynamic advanced information and communication technology-based systems that are so central to all aspects of industrial and social change.

These technologies, and their associated networks and information services, are expected to increase the flexibility of production, to contribute to environmental protection, and to safeguard democratic processes. In spite of the asymmetries in their development and use, the prospect of accessible, easy to use, interactive communication networks is consistently associated with beneficial social and economic transformations. Applications in support of education, health services, public sector information services, environmental protection, teleworking and faster, leaner, just-in-time design and manufacturing processes are also expected to transform the qualitative, as well as the quantitative, experience of productive employment and leisure time. This favourable scenario is found both persistently and insistently in the rhetoric of the producers and advertisers who promote the use and application of growing quantities of electronic information.

The consequence of all this activity is the general expectation that the diffusion of information and communication technologies will bring consumers greater choice and control over their lives.[1] The economy is expected to benefit from growth stimulated by their wider diffusion and use. Even when system frictions and the need for structural adjustment are forecast in the face of the diffusion of these technologies, optimism continues to fuel the belief that a full-fledged equitable information society is only a matter of increased investment and continuous innovation.[2] A common view of the optimists is that the smooth transition to a new 'information society' or 'information economy' is merely a matter of appropriate policy and regulatory intervention. In most cases, on the other hand, policy intervention by governments is only deemed appropriate when it targets specific adjustment problem areas such as R&D expenditure, competitiveness, or training programmes.[3]

When these technologies are properly seen to be located in the wider context of their production and use, however, it becomes increasingly clear that they may play a role in disenfranchising people from their right to gain access to, and apply, information essential to their leisure and working lives; that they may completely transform deeply culturally embedded notions of individual privacy; and that they may exclude people from social and economic activity if they are unable to access or to use the networks and services available to others. As a European Union report aimed at stimulating demand for information and communication technologies has observed, 'the main risk lies in the creation of a two-tier society of have and have-nots, in which only a part of the population has access to the new technology, is comfortable using it, and can fully enjoy its benefits' (Group of Prominent Persons 1994: 6). In both the USA, where the government's 'Agenda for Action' aims at stimulating the construction of a National Information Infrastructure (United States 1993*a, b*) and in Japan with its vision of an 'infocommunications infrastructure' (Telecommunications Council 1994), particular attention has been given to the potentially divisive impact of information and communication technologies if they are used by, and useful only to, the few, rather than the many.

Are scenarios that forecast an increasingly divisive impact of the diffusion of advanced information and communication technologies simply the dystopian visions of a relatively few latter-day 'Luddites'? Or are the optimistic and negative expectations reflections of differentiated experiences of technical and institutional change? What are the fault-lines that explain these divergent experiences and why does research tend to perpetuate such divergent expectations? We argue that the fault-lines can be explained, in part, by the application of inappropriate or misleading analytical categories. These categories bias the kinds of questions that are asked about the factors that influence the innovation process and this is especially evident in research on production and use of information and communication technologies. The perception of the 'inevitability' of the trajectories of technical change results from the application of analytical categories that are unable to take into account the complexity of the human choices and actions that inform and shape these trajectories through time.

Christopher Freeman and Carlota Perez argue that advanced information and communication technologies are implicated in a major change in 'techno-economic paradigm', a change that will leave no aspect of society untouched (Freeman 1982, 1994*a*; Freeman and Perez 1988; Freeman and Soete 1994). Changes in the structure and workings of all social, economic, and political organizations are expected to be radical. And, as Perez (1983) argues, the evolution of technologies and institutions may take on many different hues; 'the final turn the structure will take, from the wide range of the possible, and the time span within which the transformation is effected to permit a new expansionary phase will, however, ultimately depend upon the interests, actions, lucidity and relative strength of the social forces at play' (Freeman and Soete 1994: 34, quoting Carlota Perez). This observation is central. It provides the touchstone for our own arguments which therefore above all concern the ways in which the choices of actors within their various institutions and environments are affecting the trajectories of change in the development, diffusion, and appropriation of information and communication technologies.

More specifically, the ways in which users become directly and actively involved in the complex process of technical and institutional change are largely ignored in the literature on the impact of information and communication technologies.[4] Technological change is often portrayed as pursuing its own 'technological logic'. For example, information and communication technologies are expected to become so user-friendly that their control is expected inevitably to come to rest in the hands of users. This view is just one manifestation of visions which regard these technologies as a 'juggernaut' that, both for better or worse, is out of control.

In contrast, we look empirically at the degree to which the trajectories of technical and institutional change are malleable in the hands of different technology producers, users, and policy-makers. In some cases, the scope for manœuvre is great; in others, it is very slight. The changes in these degrees of freedom, and the factors contributing to these changes, are central to the exercise of power by individuals, firms, and other organizations. By analysing the scope for choice in the development and implementation of information and communication technologies, we find opportunities for action that could, we believe, ameliorate some of the disruptive aspects of innovations in advanced information and communication technologies.

The authors in this volume contribute to the elaboration of conceptual categories that orientate empirical inquiry towards questions that are fundamental to an understanding of whether the development and use of information and communication technologies can be guided in ways that contribute to socially, economically, and environmentally sustainable goals. Transformations in the capabilities of producers and users of information and communication technology systems, and the contradictions that emerge as people in the public and the private spheres engage in the design of technologies and in establishing the rules of conduct that govern the innovation process, are central focal points in our analytical approach.

Our approach, therefore, involves an analysis of institutional and technical change that embraces the cultural, social, political, and the economic dimensions. It focuses on uncertainty and ambiguity, changes in institutional and technical boundaries, and the power relationships that are associated with new electronic information and communication technologies.[5] The determinants of qualitative and quantitative changes are made visible through analysis of how these technologies influence the dynamics of family and household relationships, the organization of work practices, and the effectiveness of public policy and regulatory measures.

No claim to comprehensive coverage of all the social science disciplines is made, nor do we seek the integration of theories and concepts in a search for a new overarching theory. Our enquiry moves towards a 'middle-range' theory of innovation, integrating social, political, and economic perspectives on the dynamics of technical and institutional change. A 'middle-range theory' is a way of conceptually framing issues so as to bring salient research questions to the fore.[6] Merton, for example, describes 'middle-range' theories as those applying to limited social experience. Middle-range theories are helpful in developing explanations because they group together assumptions that can be examined for their value in assisting social science inquiry (Merton 1968). This approach is also akin to the use of a 'heuristic frame' which 'corresponds to a degree of problem definition that occupies an intermediate position on the continuum between a long and indiscriminate list of things that might matter at one end and a fully formulated control-theoretic model of the problem at the other' (Winter 1987: 167).[7] Central to both formulations is a concern with the need to ground theory in careful and sensitive empirical study (see Glaser and Strauss 1968).

Our approach considers power and its differentiated articulation through human action.[8] It differs from ecological theories of technical and organizational change which have become increasingly advanced. Such theories use a game metaphor to generate formal models of decision-making. Emphasis is given to the optimization of efficiency and the aim is to interpret behaviour in the context of the games individuals are perceived to be playing. Like our approach, the models of decision-making arising from ecological theories abandon the assumption that human action is perfectly rational (Dutton 1992: 323). However, they do not take into account that the actors in the game do not have the same capabilities 'to shape their fate as they communicate with one another to define the games' (Dutton 1992: 323, citing referee's comments; Shields 1995 forthcoming).

Our middle-range approach places a 'dialectics of outcomes' at the centre of analysis in contrast to the literature on the 'politics of decision-making' which focuses mainly on processes of individual negotiation in collective settings (Danziger *et al.* 1982; Dutton and Kraemer 1985). Our concern is with the way the trajectories in the development and use of information and communication technologies are selected. We locate information and communication technology users in the home and the workplace; we investigate their activities as producers and users of information and communication-related goods and services; and we assess the way actions affect power relationships in dynamically changing institutional settings.

We argue, therefore, that the innovation process should be treated as a dialectic in which power is exercised in the production and use of technological artefacts as well as in the institutionalization of behaviour. Central to this argument is the methodological challenge to undertake research which explores the dialectical relationships at work in the innovation process whether they are worked out in the household, the firm, or the intergovernmental organization. Our approach draws on Anthony Giddens's structuration theory which offers a departure both from perspectives that privilege the individual as a rational self-maximizing agent (reductionism) and those which reify organizational (institutional) structures.[9] Giddens's approach is one among a number of social theories which have in common the assumption that agents and structures are co-determined. These theories begin from an assumption that the relationships between agents and structure are neither fixed by historical events, nor completely malleable in the hands of individual actors.[10] Innovation results, consequently, from the intended and unintended consequences of actors within and across the changing boundaries of organizations.

Such a dialectical analysis points to the working out of the contradictions between the intended and unintended consequences of human action.[11] It offers a mode of critical inquiry into otherwise untested assumptions about 'good' and 'bad' outcomes that characterize much research on innovation in information and communication technologies. It offers a technique for clarifying controversial issues by giving attention to the processes whereby contradictions are continuously resolved through time. And, finally, it suggests a way of understanding how qualitative and quantitative transformations in the power relationships in society are associated with technical change.

We draw upon our own empirical studies to illustrate biases in institutionalized relationships, or the structural privileges, that are maintained or created anew as information and communication technologies are developed and diffused. The empirical work was undertaken from 1986 to 1995 and our perspectives emerged out of the continuity of exchanges among a multidisciplinary group of scholars.[12] A workshop debate prior to the launch of our research programme pointed to the importance of the analysis of the degrees of freedom open to social actors who—whether by design or default—alter the trajectories of information and communication technologies.

The examination of a society's communications system reveals that it concerns more than decisions about spectrum allocation, digital transmission and switching, fibre optics, cable systems, deregulated telecommunications or geostationary orbit positions. Such decisions represent more than technical, market or organizational decisions about *means*. They are also political decisions about *ends*, and the particular social objectives—industrial, cultural or military—which communications systems contain and mediate. The public policies which mould communications systems are also social policies—by design or by default. (Ferguson 1986: 66–7)

We show that the design of information and communication technologies is the result of contradictions in actors' choices as they become embedded in relatively stable institutions. The negotiations which enable technology producers and users to design and appropriate information and communication technology systems in ways that reproduce or change patterns of behaviour are complex and varied. They are understood in our work as being the result of the intended and unintended outcomes of action by individual consumers, as well as by public policy-makers and corporate actors. Thus, innovations in information and communication technologies are characterized by varying degrees of openness and closure, and by acceptance and resistance. The innovation process itself is shown to be more complex than can be acknowledged by most formal models.

The spheres of influence or 'careers' of a wide range of information and communication technologies are considered (Silverstone 1994*a*). In some instances, the technologies we consider are at an early stage in their development life-cycles. In others, they are deeply embedded in households, firms, and public organizations. We focus mainly on advanced information and communication technologies because these are widely believed to hold great potential for interactivity in the knowledge creation and application process.[13]

Advanced information and communication technologies are especially complex because they are both machines and media. As such they are both the objects and facilitators of consumption. They carry both functional and symbolic significance; and, as such, they are characterized by what Roger Silverstone calls a double articulation.[14] In addition, they have implications for the physical and symbolic organization of space and time and this, in turn, affects all producer and user communities.

How then do actors in the information and communication technology supplier, user, and policy-making communities create the conditions for distinctive types of electronically mediated interaction in local, national, regional, and increasingly global communities? This question is at the core of our analysis in the following chapters. In Chapter 1, Robin Mansell lays the foundation for the analysis by outlining the conceptual approach which led us to title this book—Communication by Design. Chapter 1 takes the reader on a multidisciplinary tour of analytical perspectives to show that many of those working in the traditions of economics, political science, and sociology have been drawn increasingly to focus on communication, the capabilities of actors, and on the intended and unintended action in the design of information and communication technologies.

Roger Silverstone and Leslie Haddon begin the analysis of the implications of information and communication technologies in Chapter 2 by examining the interface between design and domestication. The latter term describes the process whereby advanced information and communication technologies are appropriated by users through their consumption. They highlight the activities of consumers who cultivate information and communication technologies as they enter the household. Domestication implies a politics of meaning and practice which engages consumers throughout the 'careers' or life-cycles of these technologies. It implies a continuous redefinition of the boundaries of the public and private spheres and constant transformation in both technologies and their users. The dialectical character of the innovation process is shown to be constituted by the interrelationships between product design, the construction of technologies in the rhetoric of public policy and marketing, and the capacity of households to define and redefine their own terms in use. Users, themselves, become participants in the process of technological change; they are not simply recipients of given technical systems.

A dialectic is visible too in the development processes resulting in software production. In Chapter 3, Paul Quintas explores the development–use continuum to locate human agents, design processes, and technical change within the context of factors in the professional software developer and user communities. His analysis highlights the characteristics of software technology pointing to its heterogeneity as well as to the uniqueness of an industrial structure in which technology user organizations are also innovators. The chapter explains the flexibility of software and the continuous process of development that is enshrined in its maintenance. Here, the user is both a recipient of software technology and a participant in the development process. These characteristics of software are contrasted with the constraints imposed by initial design decisions and by past investment in software applications which limit the degrees of freedom available for developers or users to introduce change.

Software supports the information processing and control capabilities embedded in advanced information and communication networks. In Chapter 4, Robin Mansell looks at the experiences of firms involved in the design and use of networks and services. The relationships among innovating clusters of geographically distant firms are examined to illustrate how producer–user relationships unfold. The analysis shows that these relationships are influenced both by technical network configurations and by processes of communication embodying the changing and unevenly distributed capabilities embodied in network designers and users. Firms find their choices about networks and service alternatives constrained by the characteristics of existing and new technologies, but these technologies also create opportunities for distinctive, new network configurations.

In Chapter 5, the degrees of freedom available to users as a result of alterations in the software or intelligence embedded in electronic networks are examined. Rohan Samarajiva takes up the implications of these networks for privacy in human interaction. The design of the public telecommunication network enables it to collect and store massive amounts of information as well as to create new 'public spaces' within which human interaction occurs. The electronic surveillance environment and the implications of asymmetries in producer and user capacities to collect, store, and process information are considered together with the options available to policy-makers and regulators to address problems created by changing norms and acceptable practices which alter the privacy of social and economic life.

Standardization offers a degree of control necessary to achieve interoperability and interactivity within an advanced information and communication technology-based network. In Chapter 6, Richard Hawkins shows how the technical aspects of standards and the standardization process are located in the matrix of social, political, and economic relationships which generate them. Standards provide a degree of stability in the construction and maintenance of complex technical systems. In such systems, no single political, economic, or social institution determines the nature of technology production or consumption. The power relationships among participants result in considerable indeterminacy in the design or architecture of information and communication technologies. This occurs throughout the life-cycle of the technologies and standards are shown to embody political and commercial biases or preferences that become embedded in successive generations of information and communication technology-based networks.

The governance processes and the rhetoric of policy-making that shape information and communication technologies are taken up in the next chapter. In Chapter 7, Robin Mansell looks at how users and producers articulate their requirements in ways that alter the priorities and practices of international governance institutions in the tele-communication regulatory and intellectual property rights domains. Tensions, and sometimes unexpected institutional developments, are shown to arise out of the shifting alignments among technology producers and users. The emergence of new institutions and the maintenance of older ones is shown to create opportunities as well as barriers for the diffusion of information and communication technologies and for the creative production and consumption of electronically networked knowledge.

In the final chapter, Roger Silverstone and Robin Mansell draw the strands of our analysis together. They crystallize the main aspects of 'middle-range' approaches to innovation in information and communication technologies. The specificity of innovation in information and communication technologies and its implications for the trajectories of technical and institutional change are considered. We consider whether we are on the threshold of a new knowledge society in which information and communication technologies contribute positively to social and commercial endeavours. A multidisciplinary politics, economics, and sociology of the careers of these technologies challenges us to consider this issue. We find that the answer lies in an understanding of the history of technologies, institutions, and their environments, and in the dynamically changing capabilities of actors within their complex environments. As Raymond Williams (1983: 268–9) suggests, 'if there are no easy answers there are still available and discoverable hard answers' and some of these are contained in the following chapters. Many more will be needed as information and communication technologies become more pervasively present in the twenty-first century.

NOTES

1. Control refers to increasing the probability of a desired outcome rather than to the absolute determination of an outcome. This usage follows Beniger (1986) and is discussed by Samarajiva (1994).
2. Sylvia Ostry refers to 'systems frictions' in the context of the boundary between trade policy and domestic economic policy. As Richard Lipsey argues, 'today, with services, investment, and other matters looming large, these different systems impinge in major ways on international trading and investment relations. Thus, different domestic systems come into conflict and strong pressures are exerted either to harmonize them or to manage the trade that is affected by them' (Lipsey 1994: 25).
3. National technology policy initiatives such as the Technology Foresight Programme in the United Kingdom and the European Commission's Framework Programmes tend to be characterized by actions that tackle segmented problem areas deemed to require initiatives to stimulate structural adjustment.
4. The importance of user innovations and user–producer linkages is increasingly recognized in the economics of technical change literature focusing on systems of innovation, see e.g., Lundvall (1992*a*); and in research on the social and economic impact of information and communication technologies, see e.g. Mansell (1994) and Quintas (1993).
5. See Davis (1994: 1, 24) and Gumpert (1992). Davis argues that 'the most recent developments in telecommunication based on digital technology have created a new environment of convergence and interactivity. These technologies have furthered the opportunities for human role ambiguity as well as message intent equivocality by removing some of the traditional control mechanisms that served as signposts for observers of and participants in the communication process . . . New media have created a path for unconventionality, and therefore unpredictability, in the appropriateness of shared public messages.' This view echoes those of Innis and McLuhan both of whom associated institutional instability and uncertainty with innovations in information and communication technologies and their widespread and uneven diffusion and use. See e.g., Innis (1950, 1951) and McLuhan (1951*a,b*, 1964) who argue from different perspectives.
6. For a discussion of 'middle-range theory' and its differentiation from 'meta-

theories' which may attempt to provide all-encompassing explanations, see Orlikowski (1992) and Orlikowski and Robey (1991: 165).

7. The term, 'heuristic frame', refers to 'a collection of possible approaches to a particular strategic problem whose members are related by the fact that they all rely on the same conception of the state variables and controls that are considered central to the problem' (Winter 1987: 167).

8. This approach has close affinities to the 'new' institutionalism in economics and concrete theory in political science, see Lane (1990).

9. Structuration theory represents an effort to reconceptualize the divisions between the objective and subjective that pervade much social science theory. For Giddens, 'structuration theory is based on the premise that this dualism has to be reconceptualized as a duality—the duality of structure . . . The structural properties of social systems exist only in so far as forms of social conduct are reproduced chronically across time and space. The structuration of institutions can be understood in terms of how it comes about that social activities become "stretched" across wide spans of time-space' (Giddens 1984: pp. xx–xxi). In structuration theory, ' "structure" is regarded as rules and resources recursively implicated in social production; institutionalized features of social systems have structural properties in the sense that relationships are stabilized across time and space' (Giddens 1984: p. xxxi).

10. According to Wendt, research programmes rooted in structuration theory embody, in contrast to individualist programmes, an acceptance of the reality and explanatory importance of irreducible and potentially unobservable social structures; in contrast to structuralist programmes, an opposition to functionalism and the need for a theory of practical reason and consciousness that accounts for human intentionality and motivation. Although interpretations of Giddens's structuration theory vary, it provides a basis for regarding agents and structures as being joined in a dialectical synthesis. Neither is considered to be subordinate to the other (Wendt 1987: 356). Others working in the structuration tradition include Bhaskar (1989), Bourdieu (1977), Carlsnaes (1992), Dawe (1979), Gregory (1981), Gregory and Urry (1985), Knorr-Cetina and Cicourel (1981), Layder (1981), Lloyd (1986), Mouzelis (1980), O'Neill (1973), Rosenau (1986, 1988), and Thompson (1978). Our approach assumes that the relative subordination of structures and agents varies through time.

11. The dialectic view had its origins in ancient Chinese philosophy and religion. In the 6th century BC, the Milesian school in Greece proposed that 'becoming' characterized the continuous flow and change of all things (Smythe 1991). Change arises from the interaction of opposites; each pair of opposites comprises a unity which contains and transcends opposing forces. This view emerged in Western philosophy in the writings of Hegel and Marx. The dialectic is explained by Herbert Marcuse who emphasizes negation; 'the power of negative thinking is the driving power of dialectical thought' (Marcuse 1960: p. viii). Majone suggests that 'the starting point of a dialectic argument is not abstract assumptions but points of view already present in the community; its conclusion is not a formal proof, but a shared understanding of the issue under discussion.' (Majone 1989: 6).

12. Research by the authors, except Dr Rohan Samarajiva, was sponsored by the United Kingdom Economic and Social Research Council Programme on Information and Communication Technologies (PICT) and other public and private sector sponsors. PICT funded research at five other university sites: Centre for Communication and Information Studies (CCIS), University of Westminster; Centre for Research on Innovation, Culture and Technology (CRICT), Brunel University; Centre for Research on the Management of Technical Change (CROMTEC), University of Manchester Institute of Science and Technology; Centre for Urban and Regional Development Studies (CURDS), University of Newcastle; and the Research Centre for Social Science, The University of Edinburgh.
13. Varying degrees of interactivity are enabled by different advanced ICT capabilities. Interactivity must be treated as a relative concept, see Thomas and Miles (1990).
14. See Silverstone and Hirsch (1992a: 9) where the 'double articulation' of technology is explained in the case of television—'it is an object of consumption and it facilitates consumption in its circulation of public meanings'. See also Silverstone *et al.* (1992: 28) where 'double articulation' refers 'to the ways in which information and communication technologies, uniquely, are the means (the media) whereby public and private meanings are mutually negotiated; as well as being the products themselves, through consumption, of such negotiations of meaning'.

1

Communication by Design?

ROBIN MANSELL

[A] designer concentrating on satisfying one kind of objective in detail can overlook modifications that in retrospect prove to be disastrous to the . . . machine, structure, or process that is the primary purpose of the design.

(Petroski 1994: 16)

INTRODUCTION

Changes in advanced information and communication technologies have fundamental implications for the way we communicate, produce, and use information, and accumulate knowledge. Although there can be little doubt of this, the investigative tools and concepts that would enable us to uncover the dynamics of such changes have proven elusive to disciplinary-based scholars in the social sciences. Our aim in this chapter is to outline the theoretical frameworks that underpin our investigations of how actors in the information and communication technology supplier, user, and policy-making communities are actively participating in the innovation processes that generate the technical and institutional conditions that are contributing to pervasive changes in industrial activities and everyday life.

We begin by acknowledging that advanced information and communication technologies are pervasive technologies; they have the potential to alter very radically the contours of the society in which we live (Freeman 1994a; Freeman and Perez 1988). We also emphasize that the analysis of the determinants of technical change cannot be divorced from the analysis of institutional change.[1] As Christopher Freeman has argued, whether technical or economic, all innovations are 'social and not natural phenomena; all of them are the result of human actions, human decisions, human expectations, human institutions' (Freeman 1992a, Freeman 1992b: 224).

The author particularly wishes to thank Professor W. Edward Steinmueller, Maastricht Economic Research Institute on Innovation and Technology, The Netherlands, for his detailed and helpful comments on an earlier draft of this chapter.

Studies of technical change and the innovation process are giving increasing emphasis to the importance of institutional change and, in consequence, the limitations of formal theories in examining the variety of deeper processes that influence the trajectories of technical, social, and economic development are becoming ever more apparent.[2] In the economics field, Richard Nelson has suggested, for example, that recent developments in formal theories raise questions about 'how far economic selection arguments can take one in an evolutionary analysis of economic change, and the extent to which political and social forces need to be taken explicitly into account' (Nelson 1994: 53). Nelson and others use the label 'appreciative theory' to describe analytical frameworks which involve theorizing, but which remain relatively close to empirical data (Nelson and Winter 1982). Such theories are regarded as yielding potentially productive insights into the complex processes of technical and institutional change and they are concerned with political and social, as well as economic forces.

While some economists are straying into the traditional disciplinary territories of sociologists and political scientists, the reverse is also happening. In the previous chapter, we argued that 'middle-range' theories, encompassing certain bounded aspects of the social and economic experience of innovation in information and communication technologies could help to shed light on the processes of technical and institutional change that are at work. Middle-range theories are explicitly partial (Merton 1968). They lay claim neither to the predictive power of formal theory, nor to the simplistic assumptions about human behaviour that are necessary attributes of formalism. Middle-range theories are therefore close cousins of appreciative theories.

An inquiry that seeks to understand innovation processes on both the technical and institutional fronts must draw upon conceptual tools that are emerging within the various disciplines of the social sciences. Christopher Freeman has suggested that changes in institutional frameworks come about for a variety of reasons.

People have ideas to improve existing institutions and to invent new institutions and are able to muster enough political support to make changes. . . . The existing institutions are unable to cope with new problems or the growth of old problems or can no longer function effectively because of changing circumstances. (Freeman 1993: 183)

The social, political, and economic dynamics of these changing circumstances, that is, the creation of the new institutionalized modes of behaviour that are mediated by electronic information and communication technologies, are central to our analysis in subsequent chapters. We position our inquiry within the terrain of economics, political science, and sociology. In the process of traversing multiple disciplinary boundaries, the reader may find that some concepts are used in ways that loosen their familiar meanings. We attempt to ensure, however, that such concepts are located in the disciplinary literatures where they find currency and to explain any new meanings that are ascribed in the process of constructing our theoretical perspectives.

CONSTRUCTING MIDDLE-RANGE THEORIES

The observation that the design and selection of technical systems are complex social processes that emerge from continuous negotiations among supplier, user, and policy-making communities is appearing with increasing frequency in the literature on technical change and innovation. The recognition that these processes do matter and that they should not be assumed away in the interests of theoretical elegance, is leading many social scientists to focus their inquiries on the changing array of perceptions and priorities that influence the behaviour of human actors. Perceptions of what constitutes an advance in technical sophistication and of the adequacy of the 'fit' between technical and institutional systems is increasingly understood as the result of the evolution of socio-economic and technical systems through both discourse and practice. Those engaged in the process are influenced in their actions by cultural values, norms, and archetypes that are constructed within physical and electronically bounded institutions and environments. The theoretical formulations of disciplinary-based scholars are becoming better attuned to these aspects of technical and institutional change as they seek to understand the determinants of these changes and their socio-economic consequences.

Scholars in the social sciences are using a number of similar concepts to inquire into the way power is exercised and embodied in human actions, institutions, and technologies. Some of these, including path dependency, learning, selection mechanisms, and trajectories, are drawn from evolutionary economics and economic history and they are appearing with increasing frequency in work within other social science disciplines.[3] Sociological perspectives on the social construction of technology and perspectives on regime formation drawn from the political science literature are also making appearances in the evolutionary and institutional economics literature. In all these areas of social science inquiry, renewed emphasis is being given to analysis of the tensions between human agency and structural transformations that occur throughout the innovation process.

Either because advanced information and communication technologies enable an increasing 'density' of communication or because of the emergence of new modes of socially distributed knowledge production that involve many producers and users of information and knowledge (Gibbons *et al.* 1994), there are clear signs that social scientists are seeking to mine each others' disciplinary territories. As the conceptual apparatus of disciplinary scholars trickles more easily through the long-established boundaries within the academy, there are also signs of growing co-operation among those working within similar epistemological or conceptual frames, despite their formal alliances with disciplinary research traditions. The questions that are being posed with respect to the determinants of technical and institutional change, and their resonances with the interests of those in the business and policy-making communities, are creating new common ground for multidisciplinary inquiry. As Diesing (1971: 11) has suggested, although formal and evolutionary or institutional economists may 'have little polite to say to each other', the latter can often work closely with sociologists. Our inquiry brings several elements of these theoretical traditions together in order to investigate the social and economic aspects of technical change in the information and communication technologies field.

In spite of these developments in multidisciplinary inquiry, much research on the determinants of the innovation process in information and communication technologies continues within the confines of disciplinary inquiry. On the one hand, for example, issues concerning the symbolic content of the media and the informational aspects of electronic networks are examined separately from those concerning industrial structure, regulation, and policy (exceptions include Downing *et al.* 1990 and Mattelart 1991). Studies of the distinctive experiential characteristics of electronically mediated communication processes continue to draw predominantly on theories of communication, network design, policy formation, education, organization, linguistics, and computer science, giving little attention to insights that might be derived from studies located in economics (Benedikt 1992; Harasim 1993; Rheingold 1993). Similarly, the ways in which symbolic meanings are attributed to the technical artefacts of communication or to the images they convey are examined mainly by those working within sociology and closely related traditions.

On the other hand, the explosive growth in communication as a result of the expansion of networks like the Internet and the development of high-capacity public communication networks are spawning new concerns about who will pay for network upgrading and how public or universal service obligations should be defined. Questions are being raised with a renewed note of urgency about how policies and regulations can be designed to create incentives for private sector investment in information highways and their extension to consumers, businesses, and public organizations. These questions, invariably are considered mainly by economists and, less often, by political scientists.

Both formal and appreciative theorists are struggling with how to reflect the complexity of all these aspects of human agency and action in the analysis of the determinants of technical change and innovation. Criticisms of formal theories which are premissed upon the hyper-rationality of human agency and upon the notion that uncontested criteria can be established to judge whether a technical innovation is 'optimal', are leading increasingly to the rejection of simple linear models of the innovation process and technical change (Freeman 1982; Nelson and Winter 1982). These models are rooted in classical Newtonian science where relatively simple systems with a limited number of variables could be analysed mathematically to solve equations linking cause and effect. Nathan Rosenberg (1994a: 139) has argued that 'everyone knows that the linear model of innovation is dead', and there are renewed attempts to embed more realistic premises about the nature of social and technical systems in the theoretical perspectives that are used to guide and inform the empirical work of social science inquiry.

In the economics literature, for example, the perfect or hyper-rationality of agents is being seriously challenged and, as Brian Arthur suggests, an important question is now 'how to construct economic models that are based on an *actual* human rationality that is bounded or limited' (Arthur 1993: 1). The concept of bounded or limited rationality is derived from Simon's (1945, 1955, 1988) work in organizational theory on the satisficing behaviour of boundedly rational human beings and from parallel observations by political scientists such as March and Simon (1967) and March and Olsen (1989). In the field of econometric modelling, it is now beginning to be recognized that agents, whether conceived as firms, government agencies, or consumers, learn as a result of experience. Such agents may change their preferences and actions as a result of many different kinds of learning experiences and this must be taken into account if models are to be strengthened. As Arthur (1993: 18) puts it 'transience may beget transience, and lack of equilibrium can percolate across the economy'. In recently developed formal models of economic growth, for example, agents are assumed to adjust continuously to changes in their environment; and in contrast to the restrictive assumptions that have been used in the past, they do not necessarily choose optimum solutions in their activities.

Economists are not the only practitioners of academic inquiry who have worked within the restrictive framework offered by formal linear models of innovation. The other social science disciplines are equally culpable. Ian Miles and Jonathan Gershuny's (1986) critique of the linear innovation models which envisaged the graceful and predictable emergence of the benefits of advanced information and communication technologies was rooted in a rejection of social theories which had been premissed upon the hyperrationality of human agents. Their appreciative theory of innovation in advanced information and communication technologies sought to reflect the complexities and uncertainties of sociotechnical trajectories of change.

The intervening decade since Miles and Gershuny published their analysis of the social economics of information technology, together with the demise of linear models of innovation and technical change at least within the academic community, has seen the birth of models that are often collectively referred to as complexity models. These models recognize that living systems are dynamic, and that they are characterized by non-linearity and an inherent instability. These insights have given considerable momentum to econometric modelling exercises in all the social science disciplines. The 'discovery' that social and technical systems are complex has also coincided with the availability of less costly, and increasingly powerful computers which are enabling the variety of human behaviour to be modelled through the incorporation of many more variables in computational algorithms.

Some of these models are being developed with the aim of improving the predictive power of formal theories of social and economic systems and of technical change. Much effort is being devoted, for example, to analysing the workings of complex adaptive biological and human systems (Arthur 1990; Ruthen 1993; Stein 1989; Waldrop 1992). One research strategy in this area involves a focus on the self-organizing capabilities of social and economic systems in order to understand the changes and transitions in human systems. This is giving rise to a new research agenda in experimental economics and this work remains largely outside the boundaries of appreciative or middle-range theories. As our interest is in the determinants and outcomes of technical change and innovation which 'depend on the social and political conflicts which accompany this transition', we do not pursue this area of research any further in this book. As Freeman argues, these transitions 'are part of a unique human historical process, not amenable to the illumination of biological evolutionary analogies. Reasoned history and sociopolitical debate remain as the essential tools of the social scientist' (Freeman 1994c: 21).

In spite of the growing permeability of disciplinary boundaries and attempts to admit the complexity of social systems into the analysis of the determinants of change in technical and institutional systems, there still is much evidence of incommensurable inter- and intra-disciplinary world views. Perhaps because of the pervasiveness of the implications of advanced information and communication technologies, and the rapidity of change in their design and use, the limitations of disciplinary inquiries are particularly evident in this area. Our middle-range perspectives are intended to create the analytical bridges for inquiries into the 'deeper processes' of technical and institutional change. The challenge which we take up is the construction of theories that draw on the strengths of perspectives which seek a deeper understanding of the dialectics of human agency and structural change.

AXIAL PRINCIPLES AND THE MATRIX OF HUMAN ACTION

The method employed here begins with the identification of axial principles. These are used heuristically to highlight similarities in the conceptual treatment of innovation and technical change that can be found in various strands of the social science literature. We encapsulate these principles in two concepts—*design* and *capabilities*. The first is used to draw attention to the importance of human agency and its complexity. The second is used to focus on the dynamics of the communication process and the formation of human capabilities through the processes of technical and institutional change.

Daniel Bell has suggested that axial principles point, not to causal empirical relationships, but to the relative centrality of issues. Axial principles are energizing principles which can provide an organizing frame for research. They offer a way of expressing the primary issues that are of concern in a conceptual scheme. For example, Bell argues that,

Many of the masters of social science used the idea of axial principles or axial structures implicitly in their formulations. . . . Conceptual prisms and axial structures are valuable because they allow one to *adopt a multi-perspective standpoint in trying to understand social change*, but they do not forgo the value of perceiving the 'primary logic' of key institutions or axial principles within a particular scheme (emphasis added). (Bell 1973: 11)

He suggests that 'new economic interdependencies and new social interactions' are being created with innovations in information and communication technologies. As new networks of social relationships are formed, 'new densities, physical and social' become 'the matrix of human action' (Bell 1973: 189). The production and use of knowledge are emphasized in Bell's work as the primary logic of the post-industrial society. This approach has been criticized for its simplistic treatment of the causal relationships governing the emergence of new technologies and institutional forms (Garnham 1994; Leiss 1976; Melody 1987; Stehr and Ericson 1992*a*). Nevertheless, Bell's designation of the 'matrix of human action' as the site where the physical and social or immaterial dimensions of change are most visible, while not a unique contribution to the social sciences, provides the touchstone for our conceptual framework.

Change becomes especially visible in periods during which there are particularly great tensions and disjunctures in power relationships and their expression in existing institutions and the environment. For example, the 'moral economy' was a phrase coined by E. P. Thompson (1971) to characterize a matrix of human action which involved periods of tension and resistance during the transformation of the pre-industrial capitalist market when it came under threat from commercial forces. During this period, intricate patterns of symbolic and statutory expectations were institutionalized in ways that altered the accumulation and exercise of power. In our context, the information society, the information economy, and the knowledge society are the expressions that are used to describe the economic and social systems in the late twentieth century (see, for example, Drucker 1969; Lane 1966; Lyon 1986; and Melody 1981). However, despite the many references in the social science literature to the trajectories of change in information and communication technologies, we need to explore more deeply the determinants of such trajectories if we are to understand how, and with what consequences, they are being forged.[4] It is widely recognized that the paradigmatic changes associated with advanced information and communication technologies are characterized by tensions and resistances and by many creative opportunities (Freeman 1982, 1994a). Yet relatively little is known about the underlying dynamics and tensions that characterize this particular matrix of human action.

The linear models of the innovation process assume that logical adaptive processes will ensure a relatively smooth working-out of the impact of changes in technologies and institutions, in the interrelationships between actors and structures, and that recurrent and new patterns of behaviour will be harmoniously established. Once this assumption is abandoned, however, theoretical approaches are needed to investigate further how technical and institutional changes are negotiated through time and under conditions of uncertainty. The principles that frame our theoretical interpretations of the determinants of changes in the development and use of information and communication technologies are outlined in the next two sections. The common theme in our inquiry is a focus on how the choices of actors in their various institutions and environments affect the trajectories of change in the development, diffusion, and appropriation of these technologies. The principles—*design* and *capabilities*—inform our multidisciplinary perspectives on the relationships among actors and their environments, and the way that these lead both to inertia and to the creation of new technical and institutional systems.

A DESIGN PRINCIPLE

The derivation of the word *design* can be traced to the fifteenth and sixteenth centuries. The word *disegno* implied a sense of purpose as well as a model and English usage continues to embody both meanings.[5] As early as 1628, the phrase *by design* was used by Thomas Hobbes to indicate any action which was conceived to have been taken on purpose or with intention. The word design therefore invokes the idea of intentionality or purpose on the part of social actors. The term sometimes is used to refer to individual self-conscious intent and, at other times, to the intentions of collective actors that can only be assumed to exist.

The meaning of design is especially significant in our work because it embodies the traits of intentionality and purpose and, therefore, of the capability to initiate, as well as to constrain, action. The word design is used in some contexts to refer to the origins of intentional human action. This usage is rooted in the 'argument from design' and in the notion that the universe manifests the divine forethought of an intelligent Creator. This argument is associated with assumptions about the hyper-rationality of human action and provides a basis in the scientific and theological literature for assertions to the effect that the ultimate constituents of the world are 'ideas, such as universals, discernible by a faculty of pure or non-sensory intuition' (Tennant 1956: 73). If the ontology of all natural and human order and disorder can be located with an intelligent Creator or within a self-organizing system, then the intentions of social actors can be presumed to result from rational, optimized actions or choices—and formal theories can maintain their integrity.

There is another meaning of design which invokes only the existence of human purpose or intentionality. In this usage, social actors are assumed to have an idea of the situation to be reached, a desire for that situation because they value it, and some means of attaining or actualizing an outcome through a process of selection (Hodgson 1989; Tennant 1956). In this sense, design can be conceived as an *active* process of marking out or indicating by a sign or symbol. It is therefore closely associated with individual and collective actions which create differentiation and variety in social systems by marking out by signs, symbols, or other boundary indicators. Differentiation, variety, and boundaries are the essence of information and of the human communication process. As Bateson (1979: 223–4) has argued, for example, 'a world of sense, organization, and communication is not conceivable without discontinuity, without threshold', and the process of designing is fundamental to the appearance of continuity or discontinuity in natural and social systems. By focusing on design in the different

contexts in which innovations in advanced information and communication technologies occur, we can begin to unravel some of the factors
which produce contradictions—continuity and discontinuity—in the
social and economic actions which constrain and enable technical and
institutional change.

The concept of design has a long association with the determinants of
social, economic, and political action. It is a multidisciplinary concept
which is often used in social science inquiry to draw attention to the
determinants of the capability to act in diverse contexts. The mechanization of design by using machines in architectural artistic craftwork led, at
the turn of the century, for example, to concern about how the use of
machines would affect all aspects of economic and social life. The use
of machines would yield work of a high standard only as people
'mastered the machine and made it a tool' (Pevsner 1936: 35–6). Later,
comprehensive anticipatory design science provided a basis for coupling
the technical aspects of planning with the subjective way in which the
human intellect 'organizes events into discrete and conceptual inter-
patternings' (Buckminster-Fuller 1971: pp. viii, xix). In the architectural
literature, design is often synonymous with purposive action: 'all that
we do, almost all the time, is design, for design is basic to all human
activity. The planning and patterning of any act towards a desired,
foreseeable end constitutes the design process' (Papanek 1971: 3).

William Morris's work epitomizes the close association of the
development of English traditions in design with changes in institutionalized political and social norms (Stansky 1985; Thompson 1955).
Morris's ideas of design had a considerable impact on the trajectory of
British architecture and the tensions and contradictions in the politics of
design were ever present in his work. For example, he expected mass
production techniques to enable the promulgation of art to the masses,
but he also found it difficult to acknowledge that these same production
techniques might displace individual crafts and create a crisis for
employment.

In his discussion of Julien Gaudet's *Éléments et Théories* of architectural
design in the early 1820s, Banham illustrates how the prevailing views of
that period resulted in the 'taken for granted' acceptance of architectural
styles or designs. He shows how 'the symmetrical disposition of the
parts of a building about one or more axes was so unquestionably
the master-discipline of academic architecture that there was no need
. . . to discuss it' (Banham 1960: 16). As in the context of architectural
design, however, there is a need to discuss the underlying contradictions in social systems if we are to understand how the creators and
users of technical designs acquire new capabilities that lead to change.

The literature on technical and institutional change in the social
sciences frequently invokes the concept of design in reference to the

need to create new institutions that may be more effective in resolving disparities, for example, that accompany economic growth and technical change. The authors of a work on *Modernization by Design*, for instance, are dedicated, by design, 'to diminishing the force of material scarcity by erecting a structure of ascetic and aesthetic demand on a material base sufficient for the health and modest well-being of all' (Morse *et al.* 1969: 381–2).

Management practices, like organizations, are frequently said to be designed. In *Future by Design*, Davidow and Malone invoke images of the perfection of a market-place where customers can express their preferences for products through a process of simultaneous—or concurrent—engineering. Here, advanced information and communication technologies are expected to ensure that 'everyone affected by design decisions becomes involved in the design process to make sure that the multiplicity of downstream needs (manufacturability, service-ability, market demand, and so on) are met' (Davidow and Malone 1992: 88). 'Design for manufacturing' is an expression that is used to encourage simultaneous product and process innovations, and more effective competitive strategies, and to inculcate new management capabilities such as vision, and information acquisition through tacit and codified knowledge, and learning (Susman 1992). Business process re-engineering, when it was first introduced by the Massachusetts Institute of Technology, focused on the design of information and communication systems and on the design of organizational information flows (Hammer and Champy 1993; Scott-Morton 1991).

Design concepts also play a significant role in studies of the processes whereby consumers select and apply technologies and the concept of design provides the basis for a variety of different theories of consumption (Lemoine 1988). It is a pervasive concept in the marketing literature as well where it is used to encourage differentiation among information and communication technology-based goods and services and to segment consumer and business markets for new products and services (Cawson *et al.* 1990). The design concept is also highly visible in research on the ways users become involved, both directly and indirectly, in the construction of technical design concepts for new technologies (Sharrock and Anderson 1991; Sharrock and Button 1991).

Concepts of organizational design are invoked in research on the determinants of an apparent shift from the traditional M-form hier-archical firm to network forms of organization (Imai and Itami 1984; Johanson and Mattsson 1994; Ohmae 1990). As a form of organization between the decentralized market and hierarchically structured firms, such network designs entail reciprocal patterns of communication and exchange, co-operation, flexibility, and the capacity to enable the effective sharing of knowledge (Freeman 1991; Powell 1990; Ronfeldt

1993). A notion of network design is also very much present in sociological research traditions that investigate networks of actors who are involved in the innovation process (Callon 1992), and design concepts are present too in the political science literature on the communication links among various kinds of policy communities (Haas 1990, 1992; Rhodes 1987; Wright 1988).

The national systems of innovation literature and studies on regional and international systems of innovation contain many references to the design of scientific and technical, as well as economic, educational, and political, institutions (Barré 1994; Lundvall 1992a; McKelvey 1991; Nelson 1993a; Saviotti 1993; Zysman 1994). Although this research considers the organizational features of the institutionalization of systems of innovation, relatively little attention has been given as yet to how the institutional design process unfolds through the continuous negotiation of power relationships and the dialectics of human action.

As Descartes' principle of rationality, the mechanization of design, and linear models of technical and institutional change, gained acceptance, these ideas imbued the notion of design with a certainty of purpose. Yet, as Giedion (1941: 877) argues, mechanization and the principle of rationality have not enabled us to 'establish the conditions of creative growth'. To do so requires that we become aware of:

what kind of growth is going on within the depths of our period. . . . Barriers between the disciplines and the fact that people are educated to become submerged and confined by their special fields have resulted in a lack of interest in methodological principles; so much so that sometimes they cannot even be spelled out . . . The influence of feeling on practical decisions is often regarded as unimportant, but it inevitably permeates and underlies all human decisions. (Giedion 1941: 877)

Feeling implies uncertainty and the need for critical inquiry into the intended and unintended outcomes of human action. Whether embedded in technical artefacts and the architectures of information and communication networks or in the information and knowledge systems that these technologies support, variations and the uncertainties of design processes are at the heart of the questions which we raise in this book. We invoke our design principle, not to suggest the need for an inquiry into the certainty of purposive human action, but rather, in the terminology of the sociologist, to 'problematise' the design of the technical artefacts and institutions that influence the development of information and communication technologies. We focus our inquiry both on the intended and unintended outcomes of human action and on the dialectical relationships between agency and structure.

The design principle also draws attention to the need for analysis of the past, present, and future intentions of actors and to the systems of institutions and environments in which they interact. This applies in the case of actors within their family contexts, software developers within organizational contexts, decision-makers within firms that use or supply technologies, and those who participate in the institutions that play a role in policy formation and implementation. The design principle encourages us to consider how human action determines the life-cycles or careers of information and communication technologies. It enables us to recognize that 'things, objects, technologies, texts, have biographies, therefore, in the same way that individuals do . . . their lives are not just a matter of change and transformation. Through those changes and transformations, in their birth, maturity and decline, they reveal the changing qualities of the shaping environments through which they pass' (Silverstone 1994a: 99). Design, both intended and unintended, embodies actions that constrain and create opportunities for the actions of proximate or distant others. In order to analyse how this design process unfolds, we invoke our second axial principle—*capabilities*.

A CAPABILITIES PRINCIPLE

Our second axial principle draws attention to the centrality of human capabilities in the development and use of technical systems and to the fact that such capabilities cannot be taken for granted in an inquiry into the social and economic implications of information and communication technologies. Like our design principle, a capabilities (or competencies) principle has strong resonances with the social science literature on technical and institutional change. Dosi *et al.* (1992: 208) have observed, for example, that 'economic theory almost always assumes economic competence. It is embedded in the implicit (and sometimes explicit) assumption that economic agents are omniscient, or at least hyper-rational. To admit that organizational/economic competence may be scarce undermines a very large body of neoclassical economic theory.' We have seen in the preceding discussion that design need not imply certainty or rationality on the part of human actors. Yet it is possible to analyse the extent to which the outcomes of human action can be said to occur *by design* when one focuses on the dynamic formation of human capabilities.

A capabilities principle is used in our middle-range theories to recognize that human beings are knowledgeable agents. However, we make no a priori assumptions about how various kinds of capabilities are acquired or whether they are consistent with any given technical or socio-economic system. In periods of rapid technical innovation and instability that are characteristic of changes in techno-economic paradigms, there often are severe discontinuities between the capabilities

of human actors in all walks of life and those they are expected to
acquire. The requisite knowledge sets are not freely given. In fact, on the
contrary, capabilities arise from diverse experiences and they are
the result of substantial investments of time and other resources.

Capabilities are not equally distributed throughout the population. As
Christopher Lasch argues, 'in the twenty-first century, equality implies
a recognition of limits, both moral and material, that finds little support
in the progressive tradition' (Lasch 1991: 532), a tradition that all too
often regards innovations in advanced information and communication
technologies and their widespread diffusion as a panacea for social,
economic, and environmental problems. The capabilities principle—
coupled with the design principle—helps to uncover the deeper
processes that shape the trajectories of advanced information and
communication technologies. Capabilities refer to very divergent
qualities and to differences in the power to reason and to act. This
principle provides a conceptual tool to open the 'black box' which so
often appears to characterize the dynamics of technical and institutional
change.

The capabilities principle appears in many guises in the economics,
political science, and sociological literatures. Our usage of the term
'capabilities' is similar to Wendt's concept of capacities. He suggests that
'the capacities and even existence of human agents are in some way
necessarily related to a social structural context—that they are *inseparable*
from human sociality' (Wendt 1987: 355). Capabilities are understood in
a relational context involving individuals and relatively stable institutional structures. Capabilities, in the sense of the power to act, can be
expressed as a result of action embodied in institutions (as organizations)
as well as by the actions of individual agents (see, for example, Guzzini
1993). Power can then be conceived as a capability that varies as a result
of the accumulation of knowledge as well as the constraints and
opportunities in an actor's environment.

The exercise of power gives rise to the selection of advanced
information and communication technologies within an existing
trajectory or to a selection process that represents incremental or radical
departures from the past. Formal and appreciative theorists who
are working within the economics tradition have been seeking to
understand situations in which actors make choices in circumstances
characterized by imperfect knowledge and limited interpretative
capabilities.[6] To the extent that human choices are rational, they are
recognized as being boundedly so.[7] These limitations are said to spring
from 'imperfect foresight and the limited computational capabilities of
the mind' (Metcalfe and Boden 1991: 709). The complexity of information systems and communication processes is beginning to be recognized
as having an important bearing on the decision-making capabilities

of actors and on institutions involved in the production and use of information (David and Foray 1994*a*; Metcalfe and Boden 1991).

These adjustments to the economists' perspective on the role of information and associated technologies in the decision-making process draw upon the earlier writings of organizational theorists such as Cyert and March (1963) as well as the work of economists such as Coase (1937) and Hayek (1945). The new aspect in this work, however, is the suggestion that, in the process of accumulating knowledge through co-operative and competitive strategies, firms construct their agenda (Metcalfe and Boden 1991: 709). Although the firm is described as an 'operator translating the skills and knowledge of the individual members into a collective competence', the 'operator' is recognized as being informed by rules and norms of communication which 'filter, transform and store knowledge in the organization' (Metcalfe and Boden 1991: 711). This perspective is giving rise to discussions about variations in technological capabilities and the nature and origins of the process of knowledge production and use. The determinants of capabilities themselves, however, continue to be shrouded in consider-able mystery. For example, the concept of an operator suggests that the power to act is located somewhere and needs only to be discovered through a process involving communication. In fact, it is the commun-ication process itself that generates new meanings, capabilities, and structures through time.[8]

A capabilities principle is present in research on the competencies of firms. For example, organizational, technical, and communicative capabilities are discussed in work on the economics of technical and institutional change (Aoki 1990). Capabilities are associated with the concept of competencies which has been defined as 'differentiated technological skills, complementary assets, and organizational routines and capacities that provide the basis for a firm's competitive capacities in a particular business' (Dosi *et al.* 1992: 179). The focus here is on the tangible and intangible resources that become internalized within the boundaries of firms. Firms acquire these resources as a result of their surveillance of the environment and their capacity to transform information into usable, codified, and tacit knowledge. In line with the work of Winter (1987), Teece (1988), and others (see Drucker 1969), Prahalad (1993) defines competencies as a combination of technology, governance processes, and collective learning, and Leonard-Barton (1992: 113) defines capabilities as the 'knowledge set that distinguishes and provides a competitive advantage'. Whereas early work on economic competencies excluded administrative competence, that is, 'how to design organizational structures and policies to enable efficient performance' (Dosi *et al.* 1992: 198), research on the corporate coherence of firms has begun to stress knowledge and learning processes as social, collective phenomena (Senker 1996 forthcoming), phenomena that we embrace within our capabilities principle.

David and Foray (1994a: 34) suggest that useful codified knowledge is as much a question of the 'capabilities that intelligent agents possess for interpreting and manipulating it [knowledge] as it is of the facilities for locating and retrieving what has been stored'. They refer to the 'distribution power' of systems of innovation; that is, to their 'capability to ensure timely access by innovators to the relevant stocks of knowledge' (David and Foray 1994a: 7). However, although they recognize the importance of the capabilities of agents, a conceptual framework that encompasses both the technical infrastructure of communication and the services that enable the mediation of knowledge in electronic environments has yet to be developed within the context of this work.

The foregoing contributions to a capabilities principle are drawn both from neoclassical and evolutionary traditions in economics. The institutionalist economics tradition also provides insights into the determinants of capabilities.[9] Veblen and Commons, for example, treated technology and actors' preferences as endogenous factors in economic development and both were concerned with the way agents learn and act through time (Gough 1994). Much of the work of the early institutionalists in the USA was characterized by a form of structural or technological determinism which often camouflaged their insights into the active participation of human agents in the process of technical and institutional change (Langlois 1989; Leathers 1989; Rutherford 1989, 1994; Samuels 1990). For example, Veblen (1919: 32–3) argued that, 'the material exigencies of the state of industry are unavoidable, and in great part unbending; and the economic conditions which follow immediately from these exigencies imposed by the ways and means of industry are only less uncompromising than the mechanical facts of industry itself.' However, he also felt it necessary to attack the assumptions underpinning neoclassical economics. Using the familiar phrase, 'lightening calculator of pleasure and pains', Veblen argued that people could not be assumed to behave in a rational way and that their unfolding capabilities should be of central interest. An individual was not to be regarded simply as a 'bundle of desires that are to be saturated by being placed in the path of the forces of the environment, but rather a coherent structure of propensities and habits which seeks realization and expression in an unfolding activity' (Veblen 1898: 74).

Capabilities, unfolding within dynamic technical and institutional systems, are central features in the work of many of the early institutional economists in the USA (Hodgson 1989), and this perspective finds echoes in work on the determinants of national systems of innovation in recent years (Lundvall 1992b; Nelson 1993b). In this literature, institutional innovation is regarded as being the result of 'interactive learning and collective entrepreneurship' (Lundvall 1992*b*: 9). Learning takes place in connection with routine activities in production, distribution, and consumption. Thus, for example, 'the everyday experiences of workers, production engineers, and sales representatives influence the *agenda* determining the direction of innovative efforts, and they *produce knowledge and insights* forming crucial inputs to the process of innovation' (Lundvall 1992*b*: 9).

The relevant institutions may be isolated firms, constellations of firms, or the agencies of the State. Johnson (1992: 28) suggests, for example, that institutions should be regarded as 'informational "signposts" ' and that the production of knowledge is stored, co-ordinated, transmitted and used by and through institutions. And Dalum *et al.* (1993: 311) argue that 'the *capabilities* of a firm reside not only in its machinery and in its individual employees, but also, and primarily, in its organizing capability to transform inputs into output. This capability, in turn, depends on its institutional relationships with suppliers, customers, public agencies, research institutes and on the domestic institutional set-up as a whole.' Communication processes and capabilities are important here because they result in 'collective action forming new institutions which affect both the variety creation, the reproduction and the selection of knowledge and technology' (Dalum *et al.* 1993: 301).

In spite of the growing recognition of the importance of learning (by doing, by using, by forgetting, etc.) and of the process of communication, in-depth consideration of the intermediating effects of advanced information and communication technologies and the implications for the way learning and new capabilities evolve, continues to be developed largely apart from this line of research. In this research tradition, it is primarily tacit knowledge, that is, knowledge not codified in electronic or any other form, which is of major interest to analysts of technical and institutional change.[10] Lundvall (1992*c*: 56) and others working in this tradition also recognize, however, that 'the general institutional framework—including norms and codes—represents a context for communication', suggesting an opportunity to delve more deeply into the communication processes that are mediated by advanced information and communication technologies.

We turn next to the capabilities principle as it appears in selected strands of the political science and policy analysis literature. March and Olsen (1989: 165) use the term institutional capability to denote the transformative characteristics of actors within dynamically changing political institutions. Ruggie (1993), an influential proponent of the International Relations discourse, has suggested that, like their counterparts in economics who have concerned themselves with formal theories, International Relations theory lacks an adequate vocabulary to comprehend the emergence of new systems of rule or governance.[11] Chastising grand theories for discounting the material environment and the constraints and opportunities within which social actors interact, he suggests that the study of institutional transformations must be 'predicated on the need to endogenize the origins of structures and preferences' (Ruggie 1993: 171). Krasner (1991) also suggests that the essential focus of research on changing institutional regimes should be on the underlying distribution of technical *capabilities* among States. Technical capabilities, for instance, in information and communication technologies, are understood simply to be conferred by new technologies, and few efforts have been made to investigate the specific ways in which advanced information and communication technologies are implicated in political power relationships.

Wendt is among the political scientists who bring social interaction and the capabilities of actors into the centre of their analyses of institutional change. Wendt defines an institution as:

a relatively stable set or 'structure' of identities and interests. Such structures are often codified in formal rules and norms, but these have motivational force only in virtue of actors' socialization to and participation in collective knowledge. Institutions are fundamentally cognitive entities that do not exist apart from actors' ideas about how the world works. (Wendt 1992: 399)

Even if institutions are socially constructed, however, this does not imply that they are necessarily malleable as a result of individual action. Instead, the research agenda in this area calls for an investigation of constraints, contradictions, and the relative power of agents. Thus, actors' understandings and expectations 'may have a self-perpetuating quality, constituting path dependencies that new ideas about self and other must transcend' (Wendt 1992: 411).[12]

Concepts of learning and capabilities are also used in the political science literature to explain how actors alter their perceptions and actions through time (Haas 1964, 1990). For example, research on epistemic communities which influence decision-making in intergovernmental institutions recognizes that human agency lies 'at the interstices between systemic conditions, knowledge, and national actions' (Haas 1992: 2). Here, the role played by networks of knowledge-based experts in 'articulating the cause-and-effect relationships of complex problems' (Haas 1992: 2) is important to an understanding of the relationships between international structures (institutions) and human capabilities or volition.[13] Adler and Haas (1992: 385) suggest that the issues which become salient for international policy-makers arise out of communicative action and the capabilities of actors. Any specification of such power relationships must include the scope and domain in which such power is exercised (Baldwin 1980: 497). Power is treated here as a relational concept and the capabilities of actors to exercise power are treated as being context-specific and variable through time.

These research traditions in political science are concerned primarily with institutional change rather than with the interdependency of technical and institutional change. There are some within the political science community, however, who recognize that an understanding of technical and institutional change requires a 'structured and dynamic view of larger wholes' (Cox 1981: 125). Cox applies concepts that include material capabilities, including technological resources, institutions, and ideas that interact to produce socio-economic power relationships and institutional arrangements (Cox 1987). Nevertheless, despite his concerns with both technical and institutional change, Cox's contributions to the International Political Economy tradition are not intended to explain how new institutions emerge or why they so often appear to stagnate in the face of widespread change in the technological 'material' system (Comor 1994), the theme which is central to our inquiry.

In the policy analysis literature there is evidence of attempts to grapple with the uncertainty and complexity of organizational change by investigating the capabilities of actors, how they come into being, and the consequences for technical and institutional change.[14] As Dobuzinskis (1992: 368) has suggested, within any given system, 'individual actors who constitute . . . communicative networks exercise some measure of choice, and thus power, over their lives.' Early conceptual frameworks focusing on networks of relationships often differentiated among alternative organizational models on the basis of function. However, more recent work has turned to relational, processual, and structural interrelationships, and to investigations of the interpenetration of network domains as individual and collective actors' identities change through time (see Callon 1987, 1992; Law and Bijker 1992; and Strauss 1982, 1984).

In the sociology tradition, constructivist approaches have been developed to deepen understanding of the histories of the capabilities that guide technical and institutional change (Bijker and Law 1992; Callon 1992; Callon and Law 1982). The research agenda of those who are often collectively grouped under the social constructivist label is differentiated and fraught with its own internal arguments. These approaches tend to regard innovations in information and communication technologies 'neither as the outcome of the logic of technological reality, nor as the reflection of the reality of contextual variables' (Bloomfield *et al.* 1994: 145). Together with other contributions from sociology, this perspective has succeeded in drawing attention to the way that the social histories and biographies of individual and collective actors affect the entire life-cycle of technologies from their conception to their abandonment (and even beyond from an environmental point of view). Such histories shape actors' capabilities and their selection of technical and non-technical systems. The communication process via different modes of communication, that is, various architectures of information and communication technologies, affects the histories of technologies and social relationships. In the 1950s these inter-dependencies were of interest to those concerned with statistical theories of electronic communication (see, for example, Wiener 1950). However, early progress in understanding the interdependency of technical design and the design of social relationships was overtaken to a great extent by attempts to formalize these relationships.

In the 1990s, research agenda from economics to political science and sociology are engaging much more persistently in a search for explanations of the formation of consensual knowledge, common beliefs, and learning, and how these affect technical and socio-economic environments (Stehr and Ericson 1992b). In each of these fields of work there is consideration of the way in which human and technical capabilities emerge from relationships characterized by individual histories and the enduring norms and rules of conduct that are embedded in social, economic, and political structures.

Foss (1993: 142) argues that approaches focusing on competencies, capabilities, and learning processes are only in their infancy and that further elaboration 'will involve understanding how firm-specific human capital is embedded in and interacts with the social capital (organizational capital) of the firm'. Freeman (1992c: 181) similarly suggests that 'the *capacity to communicate and interact* with a variety of external agencies' (emphasis added) is one of the main ingredients of successful innovation. However, these observations provide only a starting-point for an inquiry into what such capacities entail and how they evolve.

To make progress in such an inquiry, we argue that an improved understanding of the process of knowledge generation and use is needed. One increasingly important aspect of this process is the particular way in which knowledge is generated and used in electron-ically mediated relationships. We have seen that the design and capabilities principles are present in various guises throughout the multidisciplinary research agenda of the 1990s. Our middle-range theories use these principles to delve more deeply into the determinants of the behaviour of actors which generates the new technologies of information and communication and influences their use. In the next section, we offer a profile of these technologies as a context for our inquiry in the following chapters.

THE TECHNOLOGIES OF INFORMATION AND COMMUNICATION

The information and communication technologies with which we are concerned include telecommunication and computing hardware and software, audio-visual equipment, and the information content of these technical systems. The sophisticated computational capacities of today's advanced information and communication technologies can be offered on a stand-alone or on a networked basis. In both cases, the intelligence or software embedded in these technologies is critical to system performance. CD-ROMs and CD-i players are mainly developed for use on a stand-alone basis, but, increasingly, information storage and processing devices are being linked together by communication networks. Although the strength of demand varies considerably across industrial sectors, socio-economic categories, and the territories of countries and regions, video conferencing, interactive multimedia, virtual reality, computer-aided design and manufacturing systems, and many more services and applications are generating growing demand for the expansion and upgrading of telecommunication networks.

Intra-organizational networks such as local area networks are being interlinked with corporate wide area networks and with the public telecommunication networks which reach into a high percentage of households and businesses in most industrialized countries. These networks, in turn, provide access to a wide array of information and communication services. Over the past decade the trend has been towards greater interworking and interconnectivity among disparate networks and services which support data, voice, and image communication (Noam 1994*a*, *b*). This networking phenomenon, whether on a local, national, or international basis, is challenging the viability of existing industrial structures, organizational norms, and rules of conduct. It is bringing disparate cultural, social, and economic phenomena together across space and time in ways that we are only beginning to comprehend.

On-line databases, electronic publishing, and multimedia services, as well as voice telephony services such as call return or calling line identification, are bringing new and difficult policy issues to the fore in countries around the world. In the communication policy community these issues are addressed in the context of rules and regulations on ownership and control of communication networks, cross-media ownership, and the limitations of existing intellectual property rights regimes (Egan 1994; House of Commons Trade and Industry Committee 1994). In the computing community, it is the Cyberpunks and the navigators of the World Wide Web and other Internet services who are capturing the attention of researchers, industrialists, and policy-makers (Sterling 1992).

Until recently, these disparate communities of actors have had relatively little in common, but it is now becoming evident that traditional operators of the communication network infrastructure, including the telephone companies and the cable companies as well as the broadcast entertainment industry, are being challenged by companies who intend to market an enormous number of information and communication services over the Internet. This network grew out of the need for data communication in support of military, scientific, and educational activities, and although it uses the underlying telecommunication infrastructure, it is becoming the site of substantial commercial information-related activity which is offered outside the framework of the services provided by traditional suppliers of communication services. Many of the visions of the future information 'highways' foresee the ultimate integration of all these networks to forge a virtual 'system of systems' that will become pervasive in the social and economic order of the twenty-first century.

In the face of the competition between cable television and public telecommunication companies as well as satellite-based suppliers of entertainment, educational and business orientated programming, 'interactivity' via electronic media of the 1990s is acquiring a new salience. Services like dial-up-video, telebanking, and teleshopping are whetting the appetites of existing and would-be information suppliers. Marketing and advertising companies project that demand for these and other services eventually will become enormous. However, these same agencies admit to substantial uncertainty as to when this market will really expand, how people will be expected to communicate, and how they will interact.

Studies by Miles and Gershuny in the mid-1980s led them to observe that 'any process of technical change unevenly benefits people at different locations in a structure of social inequality' (Miles and Gershuny 1986: 34). They were concerned with whether innovations in information and communication technologies would offer the potential for greater numbers of people to produce and use information more effectively as a result of their interaction with electronic communication networks. They suggested that the design of networks and services would have far-reaching implications for the capacity of people to develop and use them in socially and economically productive ways. As earlier observers of the history of changes in information and communication technologies had argued (Machlup 1962; Wiener 1950), they concluded that the potential for the creation of new social and economic inequalities as a result of biases in the technical designs of the new systems would continue to exist.

The legacies of historical investment in infrastructure and the interests of those who stand to gain financially from these markets create biases in the trajectories of technical and institutional change. Information and communication technologies are no different from other technologies in this respect. There is, on the one hand, a view that technical innovations and investment in new terrestrial and radio-based broadband networks will eradicate inequalities in the accessibility of today's networks and services. On the other hand, another view suggests that investment in communication infrastructure and in new services will exacerbate existing inequalities.

The potential for bias in new generations of technical systems often is camouflaged by the rhetoric of the supply side of the industry as well as by some segments of the user community. This rhetoric creates the impression that information and communication technologies and services are (or soon will be) ubiquitously available to everybody. For example, AT&T has claimed that some 5.4 billion people can interact with the telephone network (Strassmann 1994). While telephone service can be accessed in every country in the world, we also need to recall that millions of people continue to live more than one day's walk away from the nearest telephone. This advertising message is only one of the many similar claims that swirl about in the bid to stimulate demand for advanced information and communication technologies and services, the supply of which is evermore finely honed to selected businesses and consumers (Gandy 1993).

The diffusion of now familiar information and communication technologies such as the telephone, the facsimile machine, the personal computer, and the television set gives an impression of historically accumulated disparities in their penetration. Diffusion has been greatest in the OECD countries, although some of the South East Asian countries have been catching up quickly. For example, in 1992 there were 409 million telephone lines in the OECD countries representing 71 per cent of all the connections in the world (International Telecommunication Union 1994). In the OECD area, the average penetration of fixed-link telephone lines per 100 population reached 47.5, and there were 5.3 fax machines per 100 telephone main lines (OECD 1994a). In the UK, 89 per cent of households had a telephone, but this level of penetration still left some 2.6 million households without a telephone (Office of Tele-communications 1994).

There were some 34 personal computers per 100 population in 1992 in the USA while the comparable figure for Europe was approximately 10 per 100 people. In the UK, personal computers had penetrated into only 15.6 per cent of businesses, residences and educational user establishments.[15] And in the UK, Italy, Germany, and France, modems which connect stand-alone computers to networks had reached into only approximately 2.0, 1.0, 3.5, and 1.0 per cent, respectively, of the households in these countries. In the USA in 1994, more than 60 per cent of households were passed by cable networks that could carry text and data services as well as entertainment; in the European Union the relevant figure was only 25 per cent (Commission of the European Communities 1993b). Clearly we are not all equally able to 'connect' into the digital age.

Mapping the uneven diffusion of advanced information and communication technologies is difficult as a result of the scarcity of data and the deficiencies of available information for international comparisons.[16] If we focus on the telecommunication sector where comparable data are available, we find that, of total world public telecommunication service revenues in 1992, North and South America accounted for 39 per cent, Europe for 36 per cent, Asia for 20 per cent, and all other countries, only for 5 per cent. The average penetration of the telephone for low-income countries was only 0.80 per 100 population as compared to high-income countries with a penetration rate of 49.14. Similarly, the world penetration of television receivers was 15.7 per 100 population in 1990, but this average conceals the disparity between, for example, the North and South American rate of 40.8 as compared to Africa where it was only 3.7. The overall penetration rate for radio receivers stood at 39.2 per 100 population, while the figures for the same regions were 98.5 and 17.1, respectively (International Telecommunication Union 1994).[17]

The details of the diffusion of information and communication technologies within countries also reveal histories of uneven rates of diffusion. The diffusion of telecommunication services in South Africa provides a graphic illustration of the linkages between these disparities and inequalities prevailing more generally in societies. In this country, telephones per 100 households reached 102.5 for the White population in 1986, while the figure for the Black population was 38.5 in urban, and 13.8 in rural, areas (Kaplan 1990).[18] In the new South Africa of the 1990s the statistical manifestation of disparities along racial lines is beginning to disappear. Telekom SA, the South African public telecommunication operator, claims it no longer compiles data on telephone penetration rates according to race as this orientation is not acceptable in the 'new South Africa' (Horwitz 1993). The linkages between economy-wide social and political transformations and the development and diffusion of information and communication technologies could not be more evident than in this country which explicitly favoured discrimination for decades.

The revenues generated by commercially provided information and communication services are significant. For example, expenditure on information services, including professional, business, and consumer information services reached ECU 10.5 bn. in 1993 in the USA as compared to ECU 3.5 bn. in the European Union (European Commission 1994). In the UK, consumer expenditure on telecommunication services accounted for 1.6 per cent of total consumer expenditure of £340 bn. in 1992. Expenditure on recreation, entertainment, and education amount to nearly 10 per cent of this total. For comparison, household expenditure on food amounted to just over 12 per cent of the total (Central Statistical Office 1994).

The information and communication technology systems of the future will be designed in the light of the historical experience of consumption patterns and the creation of new cultural, social, political, and economic preferences. These technologies will reach into some communities more rapidly than they will into others and the causes and consequences of these disparities are of central interest in our inquiry. The relatively small amount of evidence that is available to document the variable rates of diffusion for telephones, computers, television sets, and the other components of information and communication systems are only suggestive of the biases in the patterns of diffusion and use. However, the existence of an information and communication technology gap is not disputed by most observers and it is not our intention to document it further here. The origins and maintenance of the gap between the information 'rich' and the information 'poor' has economic and sociocultural dimensions. The economic dimension is articulated generally along the following lines: 'competitive suppliers of networks and services from outside Europe are increasingly active in our markets. They are convinced . . . that if Europe arrives late our suppliers of technologies and services will lack the commercial muscle to win a share of the enormous global opportunities which lie ahead' (Group of Prominent Persons 1994: 8). The social dimension is described frequently in terms of the risk of 'the creation of a two-tier society of have and have-nots, in which only a part of the population has access to the new technology' (Group of Prominent Persons 1994: 8).

Our middle-range theories and the principles of design and capabilities are used to investigate the determinants of both the social and the economic dimensions and to show how producers, users, and policy-makers participate in the development trajectories that are embodied in the statistical representations. To do so requires that we go beneath such statistical representations to inquire into the structures and processes that yield the uneven trajectories of change.

CONCLUSION

This chapter has provided the background to our middle-range theories of innovation in information and communication technologies. Our intention is not to present an overarching theory of communication or of the innovation process in information or communication technologies. This goal represents a 'holy grail' for many scholars in the communication studies field in the USA and elsewhere. However, as James Beniger has observed, in so far as there is a field of communication studies, it is concerned with 'simple linear causation, cybernetics, vulgar computer metaphors and computer modelling, and the persistently embarrassing media effects controversy' (Beniger 1993: 19). From our perspective, debates as to the appropriate scope and content of a 'field of communication research' are located in simplistic models of the impact of information and communication technologies on society. All too often, the work that is informed by these debates falls prey to all the pitfalls of the linear models (Melody and Mansell 1983) which have been cast to one side by many of those whose work we seek to build upon.

The theoretical aspirations of many of the advocates of communication studies approaches have been summarily dismissed as being 'laudably cross disciplinary, awesomely broad in intellectual sweep' (Golding and Murdock 1978: 339). These authors point out that this so-called field embraces a

staggering and often unbounded range of interests and topics. In wondering at this heterogeneity, scholars have occasionally perceived a unity which can be made secure by constructing a single theory for the field. *This view we believe to be false* . . . the primary task of mass communications research is not to explore the meanings of media messages, but to analyse the social processes through which they are constructed and interpreted and the contexts and pressures that shape and constrain these constructions (emphasis added). (Golding and Murdock 1978: 339, 352)

We would extend the reference to mass communications research to embrace studies of historical and modern information and communication technologies, but the message is clear. The key issues for inquiries into the social and economic implications of advanced information and communication technologies concern the dialectical processes of changes in socio-economic and technical systems, the dynamics of their reproduction, and how and by whom such systems are controlled.

Raymond Williams (1958: 307, 313) suggested that 'any real theory of communication is a theory of community' and that 'the actual operation of the new techniques is extremely complicated, in social terms, because of their economic bearings'. The social and economic changes in communities of actors and institutions that accompany the multiplication in the technical means of communication and information production and consumption were central to his work. And, as Dallas Smythe suggested, no satisfactory theory of communication or information will emerge until we establish a better means of analysing the dialectics of change. As he put it, 'the relation of information and communication is dialectical in nature' and it concerns social and economic relations of all kinds (Smythe 1989: 6).

In the next chapter we begin our inquiry into these social and economic determinants by focusing on the ways in which consumers domesticate information and communication technologies through the actions of their everyday lives.

NOTES

1. The term 'institution' refers to organizations 'supporting the generation, diffusion and exploitation of technological knowledge' (Dosi *et al.* 1994). We use the term more broadly to refer to relatively stable patterns of behaviour, including routines, norms, etc., that emerge as a result of communication among actors in establishing social, economic, and political relationships. See Edquist and Johnson (1995) and Rutherford (1994) for reviews of the use of the term to refer to bounded entities and to processes. Both uses, in our view, are crucial to an understanding of innovation.

2. The importance of 'deeper processes' is emphasized in Dosi *et al.* (1994: 27). Coase (1991: 73) remarks that: 'it will not have escaped the notice of some readers that this analytical scheme [the nature of the firm] can be put into mathematical form. This should give us hope but only if this analytical power is used to enlighten us about the real rather than an imaginary world. All this will require a great deal of empirical work but this is how I conceive that the basic ideas in "The Nature of the Firm" can be made a living part of economic analysis.'

3. Path dependency refers to a sequence of economic changes 'in which important influences upon the eventual outcome can be exerted by temporally remote events, including happenings dominated by chance elements rather than systemic forces . . . the dynamic process itself takes on an *essentially historical* character' (David 1986: 30). See also Dosi *et al.* (1993), Freeman (1994*b*), and Zysman (1994) for discussions of learning, selection mechanisms, and trajectories.

4. Research on technological trajectories draws upon the work of Abernathy and Utterback (1975, 1978), Dosi (1982), Freeman (1994*b*), Nelson and Winter (1977), and Sahal (1981, 1985).

5. These two meanings became differentiated in French as *dessein*, in the case of a purpose or plan, and *dessin* in the case of a design in art or architecture (Simpson and Weiner 1989).

6. See e.g. Amendola and Gaffard (1994), Dosi *et al.* (1993), and Metcalfe (1993). As Amendola and Gaffard (1994: 8) argue, the new endogenous growth models make 'no attempt to understand the working of the growth mechanism, which is assumed in the model'; and as Dosi *et al.* (1993: 15) argue with respect to some new models of learning behaviour and selection processes, 'while they embody some evolutionary account of imperfect

learning and market selection amongst heterogeneous agents they maintain stationary "fundamentals" as far as best-practice technologies are concerned.'

7. e.g. 'economic agents are knowledge processors' whose individual code 'is self-modifying and can, to various extents, adapt its structure on the basis of experience, by the internal formation of knowledge, or through the absorption of information and knowledge from other agents . . . selection reflects control of resources' (Metcalfe *et al.* 1994: 315).

8. Industry stakeholders like Xerox are recognizing that the co-production of technologies by producers and users is essential (Brown 1991). Winograd argues that technical innovation requires 'a research enterprise directed to the understanding of people and their work, not just the theory of algorithms or processors', but he also hints at the difficulties that are encountered as actors located in different settings (the corporate research laboratory, the university, consumers) attempt to negotiate developments in advanced information and communication technologies (Winograd 1991: 172). See also Woolgar (1993).

9. The old institutionalists are represented by Thorstein Veblen spanning the period from the 1890s to 1923 and John R. Commons covering the period from the 1890s to 1950. For comprehensive bibliographies of the old and the new institutionalists, see Hodgson (1989), Hodgson *et al.* (1994), and Rutherford (1989, 1994).

10. Tacit knowledge is defined as 'the knowledge of techniques, methods and designs that work . . . in certain ways and with certain consequences, even when one cannot explain exactly why' (Rosenberg 1982: 143; Senker 1996 forthcoming). Rosenberg contrasts tacit with articulated knowledge, the latter referring to that which can be formalized and codified.

11. Governance involves the exercise of power and the protection of sovereignty; the maintenance of a bureaucratic and technical apparatus, and the application of rules, both codified and tacit, in order to mediate action, see Hawkins (1994*b*).

12. Wendt (1987: 359) argues that social structures are ontologically dependent upon their components; they have 'an inherently *discursive* dimension in the sense that they are inseparable from the reasons and self-understandings that agents bring to their actions'.

13. An epistemic community is defined as 'a network of professionals with recognized expertise and competence in a particular domain and an authoritative claim to policy-relevant knowledge within that domain or issue area' (Haas 1992: 3). Haas provides an extensive bibliography of research within this tradition. See also Ruggie (1975) who argues that institutionalization should be conceived as the order through which behaviour is acted out represented by the *epistemes* or systems of social reality that are envisaged by actors.

14. See Ham and Hill (1984), Hogwood and Gunn (1984), Jasanoff (1987), and Majone (1989) for illustrations of the capabilities concept in the policy analysis literature.

15. Data provided by B. Nzekwu of Electronic Publishing Services, London. In 1992, Italy had a penetration rate for personal computers of 13%; Germany

20%, and France about 10% (Commission of the European Communities 1993*b*). This source estimates the personal computer penetration rate per 100 households in the UK at about 25 which industry analysts consider to be high.

16. See Miles (1990) and Miles and Matthews (1992) for discussions of the weaknesses of official and non-official data sources on information and communication technologies and services.

17. Total telecommunication service revenues in 1992 amounted to $US 415 bn. Low-income countries have a per capita income averaging $US 310 and GDP per capita of $US 636 or less; high-income countries have a per capita income of $US 21,005 and GDP per capita of $US 7,911 or more according to the UN definitions used by the International Telecommunication Union in 1991.

18. The Black population comprised 76% of the population, while the White population accounted only for 13%. In 1987 the White population accounted for 72% of the residential telephones in metropolitan areas; the Black population, for 9%.

2

Design and the Domestication of Information and Communication Technologies: Technical Change and Everyday Life

ROGER SILVERSTONE AND LESLIE HADDON

INTRODUCTION

Technological innovation is not just a matter of production. Consumption and use are equally essential components of the innovation process. Technological innovation is also not just a matter of engineering. Both new and old technologies are social products: they are symbolic and aesthetic as well as material and functional. Production and consumption are not related to each other in a singular or linear fashion, but are the product of a complex pattern of activities in which producers and consumer-users, as well as those who intervene in and facilitate the process of consumption, take part.

In this chapter we offer an account of the role of information and communication technologies in everyday life which focuses on innovation as a social and cultural, as well as a political and economic, process. In the first instance it takes a user's perspective, and in so doing, draws on parallel work in other fields which has recognized the role of the user in defining the meaning of texts and objects.[1] In the second instance it focuses on the question of design. We will argue that innovation involves more than merely research and development or product launch. Innovation is a process which involves both producers and consumers in a dynamic interweaving of activities which are solely determined neither by the forces of technological change nor by the eccentricities of individual choice.

We will propose a model of what we call the design/domestication interface in an attempt to make some sense of the dynamics of innovation, and in doing so we will privilege the role and perspective of the consumer. The intention is not that we should simply take a *user's* perspective on innovation as if this was a magic wand that would resolve all problems of determinacy and indeterminacy in the innovation process. Our aim is first to insert the particular characteristics of *use* into that process so as to highlight the activities of consumers who, within their distinctive and perplexing forms of rational and non-rational behaviour, both complete and rekindle the innovation cycle. And the second aim is to focus on the interrelationship of design and domestication in such a way as to identify the particular elements of the careers of information and communication technologies as they move through the spaces and times of innovation.

DESIGN AND DOMESTICATION

Design can be seen as having three interrelated dimensions, each of which is a necessary but insufficient precondition for making sense of innovation as a dynamic social process. The first dimension, and the most obvious, is that of *creating the artefact*. In this sense of design objects are fashioned functionally and aesthetically. They have to appeal and they are made to work. As Adrian Forty points out (1986: 7) these two aspects of design—the way things look and the conditions of their making—are inseparable though, as we shall argue, the conditions of their making extend beyond the activities of manufacture. The second dimension is what might be called *constructing the user*. In this sense of design, images of eventual users are incorporated into the fabric of the object, but at the same time users are designed themselves—as ideal or as necessary to complete both the function and vision embodied in the artefact. The third dimension of design involves *catching the consumer*. This places design as a central component of the wider economic and social processes of commodification and indicates the importance of recognizing both the central role that technology plays in the consuming culture of contemporary capitalism and the role of the market in defining the status and meaning of technology. Indeed, as we shall also argue, information and communication technologies, albeit in different ways, are central to this culture in two interrelated respects. They are objects to be consumed and the means—as media—for the continued stimulation of consumption.

Domestication also involves a number of different dimensions. The link between domestication and design is provided by commodification, the process through which objects and technologies emerge in a public space of exchange values and in a market-place of competing images and functional claims and counterclaims. But domestication also involves the consumer in appropriation, in taking technologies and objects home or into other private cultural spaces, and in making, or not making, them acceptable and familiar. Indeed, the process of appropriation is more than simply a matter of purchase, since what consumers do with their technologies in their homes, is increasingly important work affecting both present and future technologies. Domestication, finally, involves what we have called conversion, indicating the importance of display. It involves the various things consumers do to signal to others their participation in consumption and innovation (see Campbell 1987, 1991; McCracken 1988; Simmel 1904; Veblen 1899). These signals are not a matter only for friends and neighbours, but also for producers and their allies in the market who increasingly are involved in life-style market research which involves mapping consumption across a whole range of products and practices.

We shall therefore explore the design/domestication interface as a key to the interrelationship of industrial and social logics in the innovation of information and communication technologies, and as a way of constructing a middle-range theory of innovation which can provide a bridge between abstraction and empirical exploration. In doing so we will be drawing principally on research conducted within British households.[2] The focus on the domestic should not be seen as a special case of innovation, however, despite its particularity and its importance as the market for domestic information and communication technologies and services rapidly grows. Both the structures and the processes which we will be describing have a wider relevance, and the process of domestication, especially, should not therefore be seen as something which only takes place in the home. It should inform discussions of technological change in a wide range of institutional settings.

Design and domestication are the two sides of the innovation coin. Domestication is anticipated in design and design is completed in domestication. Both depend on a particular balance of structure and agency in which institutional processes—which are together economic, political, and cultural—both constrain and enable the capacity of consumers to define their own relationship to the technologies that are offered to, or confront, them. These constraints, which at least as far as the consumer is concerned are largely invisible, are embodied in design and marketing and in the public definitions of 'what these technologies can and should be used for'.[3] Such public definitions are variously defined in the regulatory structures governing standards or services, in the particularities of a technology's appearance and style, as well as in the rhetoric of advertising and the instructions and guidance spelled out in the manual (see Fischer 1992, on the early history of the telephone). But equally, again from the point of view of consumption, these constraints are to be found in the domestic milieu itself: in households, in peer group or neighbourhood interaction, and in the established patterns of everyday life. These will define in large degree how a particular technology will be used and, at least to a part, also the consequences of that use. The emerging character of a new technology, as well as the established character of an old one, will therefore depend on the constantly shifting relationship of actors and structures in both these domains.

CREATING THE ARTEFACT

It is a truism now to observe that technologies are more than merely machines, and that the history of their emergence is a social as much as, if not more than, a technological history. Indeed, it is a social history in so far as the production of a new technology depends upon a politics of adjustment and negotiation between engineers, entrepreneurs, managers, salesmen, experts, laymen, journalists, scientists, showmen, and users—as together they stumble their way towards the newly possible. As Carolyn Marvin (1988) has argued and convincingly demonstrated in her study of the 'electronic revolution' of the late nineteenth and early twentieth centuries, institutions take varying times to form around new machines. The history of technologies, and especially the history of the electric light, the telegraph, the telephone, and the phonograph which form the core of her study, is a history of the complexes of habits, beliefs, and procedures embedded in elaborate cultural codes of communications.[4]

One way of illustrating what might be involved in design, in the sense of creating an artefact, would be to trace the design history of a single example of an information and communication technology. Adrian Forty (1986: 200–6) has done this for radio (see Keen 1987, on video; Haddon 1988, on the home computer). Forty identifies three stages in the design history of radio. In the first stage, exemplified in the Burndept wireless receiver of 1924, the wireless appeared as a technical object, displaying in the visible array of wires, valves, and controls (or more specifically diodes, capacitors, and resistors) an extremely striking appearance and one that reflected the almost total preoccupation of both manufacturers and public with the technical properties of the apparatus. It was presented as a functional object—and, of course, in the early 1920s uniquely so. Indeed, as Forty points out, the radio provided most people with their first experience of owning a piece of 'modern' technology and as such carried a considerable weight as a symbol of scientific and technological progress.

As the rate of technical change slowed, and the radio really became an object of mass consumption, manufacturers could no longer compete with each other in terms of technical advance or advantage. Attention then shifted to the radio's appearance and the balance between functional and symbolic claims shifted towards the symbolic. Radio design entered its second stage. The problem, as Forty (1986: 202) elucidates it, for the manufacturers of the time was the production of a unique and powerful technology that combined, and had to reconcile, 'the illusory reality of broadcasting with its actual artificiality'. The purchase of a radio receiver was not like the purchase of an electric iron, the second most popular electric technology in British homes in the 1920s. Radio was a broadcast technology that linked the purchaser into a network of communication that could be both comforting and disturbing,[5] but in either case brought a distinct and other reality into listeners' homes (Frith 1983; Silverstone 1994a: 78–103). One solution to this dilemma was to put the radio into a cabinet which harmonized with the domestic furnishings and which 'at least helped to make the monstrous unreality of radio seem part of everyday life' (Forty 1986: 202; Lacey 1994; Moores 1988). However, such harmonizing also had to take into account radio's symbolic significance as an emblem of a future of progress. The resulting designs, paradigmatically those of Gordon Russell for the manufacturer Murphy, involved radio appearing in the form of modern, if not slightly futuristic, furniture. The technology was hidden in a wooden cabinet, but the cabinet was designed in such a way as to indicate its distinctive status and function when it arrived in the living-room.

The third stage involved the wedding of radio design to the image of future progress. The manufacturer Ekco, above all in the designs of the modernist architects Serge Chermayeff and Wells Coates, used the new thermoplastic material bakelite to produce futuristic designs which increasingly became, in their various subsequent transformations, the norm for electronic technologies. Forty points to the efficacy of such a design strategy: that it diverts attention away from the uncomfortable present towards an uncomplicated and appealing future, and in so doing draws a millenarian ideology into the aesthetics of the artefact.

There are a number of points which are raised by this bald history. The first is that there is a symbiotic relationship between technical, aesthetic, and sociocultural innovation. The second is that particular technologies—and especially media and information technologies—require design solutions of quite a dramatic kind, precisely because of their distinct significance as media. The third is that—and this is a theme to which we shall return on a number of occasions—these technologies, at the point at which they become objects of mass consumption, have to be designed as domestic objects, mediating in their aesthetic, the tension between the familiar and the strange, desire and unease, which all new technologies respectively embody and stimulate.

This tension is not of course the exclusive product of the new. Nor is it confined only to the technological object. In the broadest sense all technical artefacts, be they objects or services, hardware and software have to provide solutions in their design to both functional and aesthetic problems. And they have to provide in their design a resolution of the tension between the familiar and the strange.[6] The solutions adopted by successive generations of radio designers involved an attempt at what we might call pre-domestication: an anticipation in design itself of the artefact's likely place in (in this case) the home, and an attempt to offer a solution *in the design of the object itself* to the contradictions generated within the process of technical innovation. We have seen how, in the case of radio, this has been an evolutionary process. But we have also seen it as one within which a dominant design rhetoric locking technological innovation with images of scientific progress has been firmly established in the culture of the twentieth century.

Forty's account is therefore instructive in a number of ways. It sensitizes us to the complex and historically determined dynamics of the design process, alerting us to it as a rhetorical but above all as a social process. As such it offers us a first stage in our attempts to understand the design/domestication interface. But it also provides a cautionary tale for those involved in the present generation of technological change, where for example in the move from voice to video telephony (Kraut 1994), the issues will not just be those of managing technical solutions (to image or sound resolution or synchronization) but in providing a design solution which facilitates both the conversion of the telephone from one functional object (voice-to-voice communication) to another (face-to-face communication as well as video on demand) and at the same time mediates the tension between the familiar and the strange that will inevitably be associated with such a conversion. What Forty's account does not do, however, is to address the more detailed question of how the link between technological design and the user is made, nor does it discuss directly the wider set of concerns that Marvin's study addresses, which see the history of the artefact as only the tip of the iceberg of innovation.

However, there is another, though given his aims it is a legitimate, omission in Forty's design history of the radio. It is the abstraction of radio from any consideration of its status as a medium. The radio, as Raymond Williams (1974), among others, has pointed out, was not necessarily or inevitably designed for broadcasting. Indeed in its early forms, and in a subsequent, though increasingly marginal, form—in citizen's band radio—it functioned as a two-way communication device. Its appropriation by the embryonic broadcast industry (a mixture of set receiver manufacturers and government sponsors) created—by design—a media system that progressively embedded itself into the culture of the nation. This culture—suburban, white, middle class—both sustained, and was sustained by, radio's claims to inform, educate, and entertain. The emergence of a national broadcast culture, at least in the UK and in those countries where a public service network became the backbone of a broadcast system, is a major component of radio's design history. Distinct patterns of programming and scheduling were designed as broadcasters came to learn about their audiences. In that design, a distinctive national voice, what Scannell calls 'the communicative ethos' of broadcasting, was in turn created (Scannell 1989; Scannell and Cardiff 1991).

This doubling of radio as both object and medium, a feature we call its 'double articulation' (see below), is the key to the distinctiveness of media and information and communication technologies.[7] An understanding of the design/domestication interface in relation to these technologies requires that we pay attention to this double aspect, as well as to what de Sola Pool in another context, and specifically in relation to the telephone, refers to as their 'double life', drawing attention to the fact that producers are never totally in control of the ways in which their products will be used.[8] With both dimensions of doubling in mind it becomes both possible, indeed essential, to see the innovation of media and information and communication technologies as a fundamentally social process, and the particular complexity of the design/domestication interface of these technologies as one which requires an understanding of the role of both producer and user in its definition.

It is this link between design and use that we now explore more carefully.

CONSTRUCTING THE USER

Steve Woolgar (1991: 59) links the notion of design to the construction (configuration in his terms) of the user:

[T]he design and production of a new entity (a new range of micro-computers) amounts to a process of configuring its user, where 'configuring' includes defining the identity of putative users, and setting constraints upon their likely future actions.

Woolgar's research involves an ethnographic immersion into the organizational culture of a hardware producer. It also involves a conceptual requirement to consider the machine as if it were a text.[9] In this specific sense the machine (the machine as text) provides instructions for the idealized and eventual user (the two are necessarily interrelated) to 'read' the text in ways that it itself provides for and, in a sense, legitimates. What Woolgar is trying to identify is a design process through which the user is incorporated into the hardware (and software) of the machine in such a way as to enable the user's relationship to fit both with the intentions of the designer and the embodied possibilities in the functional apparatus of the machine itself (bearing in mind of course that both are disfigured by their very ambiguity). The user is configured because he or she is inscribed in this process in such a way as to be able to find in his or her dealings with the machine an 'adequate puzzle for the solution which the machine offers' (Woolgar 1991: 69).

Woolgar's concern is to establish how this design process takes place within a complex organization and to explore the determinacies and indeterminacies of boundary definition within the organization as designers negotiate with both imagined and (in this case, through usability trials) real users an acceptable set of textual characteristics for the hardware. Clearly this process is a conflictful and uncertain one. It is also the product of, and perhaps even an expression of, the particular characteristics of the organizational culture of the computer manufacturer (see Kidder 1981). Indeed, different groups involved in the design of the text-artefact have different perceptions of who the users might, or should, be and those different groups have different power within the organization to insist on their particular views being taken into account. Woolgar explores, in particular, the relations between those in the technical support and those in the marketing sections of the company.

All this is extremely important. It is clear that technical artefacts are constructed with users in mind, even if that knowledge is often tacit, contradictory, and not often tested. It is clear that the particular culture of an organization will define the particular (in any given case) resolution in the design by which, again with greater or lesser degrees of fluency, the user is configured into hardware and software products. It is necessary here to recognize that both these products are not coterminous with the object-machine and much in the way of user configuration also takes place within, and can be deciphered through, for example, the manual. It is also clear that the process of configuration is in the broad sense a political one, both in the terms which we have already identified as within the politics of the organization, and also in the relations between the company and the actual users, who are requested (required) to consult the company if the user is unable to function in the way in which he or she is configured to do within the machine-text (Woolgar 1991: 80).

But while these arguments are both suggestive and plausible, they are limited in a number of ways. Specifically they fail to clarify the relations of determinacy and indeterminacy that the machine-text is supposed to have with respect to users. The indeterminacy of the configuring process as it plays out within the organizational politics of the manufacturer turns into a kind of pseudo-determinacy when it comes to the actual relationship that the user has with the artefact. Woolgar properly insists on the provisional and arbitrary nature of the boundary that is socially defined around an artefact or technology, but this begs the precise question he is at pains to address, and to which he assumes 'configuration' provides an answer: namely, the effectiveness or otherwise of this configurational work. It is far from clear how successful these processes actually are in configuring the user.

More serious, however, is the limited notion of the user around which the whole argument is built. It may or may not be the case that, in any given organization, the user is seen in exclusively functional, instrumental, or cognitive terms. It does appear that in this case the user has just such a status, and that the usability trials, limited as they are in practice, are further constrained by a perception of the user exclusively as being at the interface of screen and keyboard. The consequent difficulty emerges from Woolgar's own apparent acceptance of that definition and in the ensuing absence of any consideration of both the machine-text and the user as part of a wider social, cultural, and economic environment. Users are not just technical users. The category mistake that the manufacturing company appears to be making may or may not have commercial consequences, but the refusal to recognize a much wider definition of the user in the analysis has profound intellectual consequences. In both cases, the user is misread and is seen as an isolated individual. And in both cases his or her status as a consumer, and therefore as someone who will engage with the technology in altogether other and more diverse ways, is denied.

It is to this wider definition of the user—the user as consumer—that we now turn.

CATCHING THE CONSUMER

Alan Cawson, Leslie Haddon, and Ian Miles (Cawson *et al.* 1995; Miles *et al.* 1992) report research that aims to tease out the ways in which firms launch new products, that is, new products which do recognizably claim to be offering something quite new technically and technologically. Such products or product areas as home automation, multimedia or messaging systems emerge as the result of a complex organizational politics. This politics is conducted in relation to a shadowy figure—the consumer—whose presence only intermittently intrudes but yet whose actions individually and collectively will determine the success or failure of a new consumer product.

Home automation and multimedia, especially, are being designed for domestic consumption. As such they have to be sold, and they have to be sold within a complex cultural space in which consumers in their various rational or irrational ways make decisions about the appropriateness or inappropriateness of a new product to their own circumstances. As we shall argue, understanding the nature of this complex cultural space requires attention to a number of different factors. Cawson, Haddon, and Miles are concerned with the questions of how innovators develop their own notions of new consumer products; how they understand the consumption processes which their products are aimed at; how, if at all, this knowledge enters into the shaping of such innovations; and what sort of knowledge, from what sources, is being drawn upon. Together these questions amount to a wider concern, which involves the design of a consumer product—in this case of course a new media or information and communication technology—and the relation of design to future use.

There are a number of strategies and tactics to be identified in this process. Innovators will draw on existing product characteristics and product trends in making their forecasts of future demand. The logic of technology, for example, in relation to speed or miniaturization—is often called upon to provide a framework for analysing future demand without any reference at all to the consumer. Similarly, and once again in the absence of, as well as a result of, any substantial consumer knowledge, the process of product launch involves the building of what Alfonso Molina (1990) would call a sociotechnical constituency. What is involved here is the mobilization of significant players across a whole industrial and commercial terrain, in such a way that the new product and the principles that drive the new product get as free a run as possible. Such sociotechnical constituencies might include groups within the producing organization, external sources of finance (both private and governmental), suppliers of complementary products, standards-setting bodies, distributors and installers, regulators and lawmakers, organized social actors, such as consumer organizations, and consumers themselves who may be involved in a form of pre-launch testing or market research, and of course the media.[10]

Alongside these activities are those in which the consumer is imagined—constructed, at least as far as the evidence that Cawson, Haddon, and Miles offer, would be too strong a word. Such imaginings might involve the intuitive stabs of individuals reflecting on their own tastes and preferences, the calling up of diffusion curves on supposedly equivalent earlier technologies, or industry lore in which stories about competitors and other products, created and fanned by the trade press and general media, circulate and recirculate.

What emerges, or should emerge, from this unstable state of affairs is what David Teece (1986) has called a design paradigm, a more or less fixed set of characteristics which emerge through the more or less simultaneous emergence of functionally equivalent and competing products, which define an integrity for a particular product in what Cawson, Haddon, and Miles (Cawson *et al.* 1995), in their turn, call a product space. A product space, in the case of, for example, multimedia, is dependent on the emergence of a number of different product configurations according to intended applications and markets—especially through hardware–software interdependence, and professional and consumer applications. One must be careful not to exaggerate the inevitability, resilience, or fixity of the design paradigm or the product space. Both are hard won and in any given case they may remain at best fuzzy, and at worst stillborn. Indeed, as Cawson *et al.* (1995) point out, the definition of the 'product space' is a continuous process which does not come to an end with the launch of the product. However, it is clear that the design and innovation process is one in which 'vision and exhortation play as critical a role as the purely technical aspects of design' (Miles *et al.* 1992: 79). Both vision and exhortation in turn depend on a successful negotiation of the politics of both organization and market. In this negotiation, what is at issue is the capture of the consumer, the potential purchaser and user, whose desires and behaviour, even for those who conduct product trials or market research, are mostly still a matter for speculation, and whose decisions and actions, both at point of sale, but just as significantly thereafter, will determine the success or otherwise of a given media and information product, and the viability of its product space.

Cawson, Haddon, and Miles (Cawson *et al.* 1995; Miles *et al.* 1992) have begun a process of investigation of the innovation and design process which extends beyond the technical aspects of the user's actual putative relationship to the machine. In so doing, they open up the question of the determinacies and indeterminacies at the heart of the innovation process, and also of design as being an element in a much more complex web of production and consumption relations. In this they extend both Forty's historical analysis and Woolgar's socio-logical one. At the same time, while this clearly is not their intention, they have yet to provide a descriptive account of the process as a whole, and especially of the intimate relations of production and consumption of a new media and information technology (but see Cockburn 1992; Cockburn and First-Dilic 1994). They also have yet to offer a conceptual framework or a theoretical perspective which advances understanding much beyond the individual case.

It is to these two aspects of the design/domestication interface that we now turn.

CD-I: A CASE-STUDY

In an illustrative case-study of the development of multimedia focusing on the early launch of Philips' CD-i (Silverstone and Haddon 1993), we sought to show how the innovation process involved a multiplicity of actors across the production–consumption divide. In particular we were concerned with how the identity of a new product like CD-i, as well as the character of the multimedia product space, was subject to competing and continuous definition and redefinition while, at the same time, the consumer-user was similarly being defined and redefined.[11]

In reviewing the findings of that research here we seek to provide an empirical bridge between the discussions on design and those on domestication that will follow. The case-study offers an account of the various elements and players that make up the multimedia story. Those players, we argue, include consumers, both imaginary and real. It also offers an account of the innovation process of a media and information technology in all its uncertainty and indeterminacy.

Much of the running in the development of consumer multimedia products has been made by the Dutch multinational, Philips. With the experience of some less than successful product launches behind them (especially Laservision) and with the expectation that multimedia would quickly attract almost all the big electronic hardware and software producers into an increasingly valuable but also competitive market-place, Philips' strategy was to establish an early foothold with what they hoped would be a commanding technology. This would, in turn, be buttressed by a number of industrial alliances and agreements on standards. It also involved continuing hardware development post-launch, particularly with a view to making full motion video available and the creation and facilitation of a software support industry, bringing together a novel convergence of video and computer technologists with different skills but little experience of collaboration. It involved, finally and most uncertainly, an attempt to position the new product in the market-place. It is this last dimension of the innovation process which provides the focus of what follows.

It is clear that finding its place in the complex and rapidly changing map of consumer electronics was going to be extremely difficult. We identified three dimensions of the problem as they appeared, at least, to the producers. The first was the problem of predicting take-up (the problem of precedence). The second was the problem of defining the product (the problem of identity). And the third was the problem of finding the consumer (the problem of the market). Together these different concerns involved questions of establishing what kind of technology CD-i was to be: whether it would for example follow the innovation/diffusion curve of CD audio, the VCR, the home computer, or the games console. It could, of course, claim links with all four. These concerns also focused on the problem of interactivity, and the distinctiveness of the new machine from what had preceded and would accompany it in the innovation process.

But the problem of identity was not confined to the status of the hardware, for it was clearly evident that the character of multimedia and its ultimate success would be determined, more than by any other single factor, by the software available. And here the decision to produce education-related software or more popular or populist software was crucial, not just in attempting to claim a market but, at the same time, in defining CD-i within a given product space. Philips' strategy, as it turned out, suffered a radical post-launch rethink, as different consumers demanding different software emerged from what had originally been incorporated into the product and the product launch.

This uncertain progress was the product of conflicting pressures within and outside the company to *design* CD-i, to design it both literally and symbolically, and to design both hardware and software. CD-i had to be defined alongside and differently from earlier generations of plausibly similar technologies, from similarly orientated product packages offered by competitors (Commodore with CDTV were indeed first to launch in the UK), and from others such as Sega and Nintendo who were following close behind. But CD-i also had to pass through the hands of advertisers and retailers, as well as across the pages of trade magazines and national newspapers on its way to the consumer. In all these cases there were cross and competing definitions as advertisers sought to define CD-i as an enhanced television set, its principal high-street retailer (at least in the UK) associated it with CD-audio, and the broadsheet journalists trumpeted a whole new dimension in home computing (Silverstone and Haddon 1993: 27–39).[12]

Through all of this, of course, the consumer-user did not exist. Most of those involved in the marketing of CD-i agreed that there was no demand for, nor understanding of, multimedia. And so if the consumer-user did not exist he or she would have to be invented. That indeed is precisely what happened and continues to happen. But this invention is not conducted in a vacuum. Feedback from early users came to magazine editors and retailers. Philips indeed conducted their own market research. The process of domestication had begun. And it had begun in design. From the design of the remote control (rather than a joystick or a keyboard) and the packaging of the machine (to look like a video rather than a computer), to the construction of the image in advertisements and at the point of sale, the public definition of CD-i was being negotiated. Of course, beyond this, such definition and redefinition continues, for with early sales came early users and early users were not, as we have already hinted, always quite what Philips had in mind.

If producers, within the terms of the present discussion, have to 'capture' the consumer, the reverse is also true, consumers have to 'find' the technology. These two processes are of course interdependent, but it is the slippage and the contradictions between them that are most instructive. The research offered insights into the complexities of the domestication process, complexities informed not just by available resources, but by household priorities; informed not just by gender but also by age and class; and informed above all by a mixture of both high and, for some, disappointed, expectations, principally with regard to the software; as well as anxieties, principally with regard to whether the new purchasers had backed the 'right' technology. Early adopters are impatient and perhaps untypical folk, and the lack of what they saw as exciting software as well as the unavailability of full motion video were the main reasons for early dissatisfaction. Equally, early adopters are individuals with clearly defined personal agenda when it comes to new technologies. Machines appeared to be bought, at least initially, for individual rather than family use. Once in the household, they embarked on a complex career as their meaning was negotiated and renegotiated through the attention and use of different family members.

Indeed, it is the various conjunctions and disjunctions between the domestication and the design of information and communication technologies which lie at the heart of the innovation process and which provide the focus of the arguments and analysis that follow. These need to be set, however, within a broader context, and can be identified as a middle-range theory of innovation (Orlikowski 1992; Orlikowski and Robey 1991). In pursuing this theory we relate our concerns to a wider debate about innovation, and draw the issues raised by this wider debate into the specific domain of the domestic and the everyday.

THE DUALITY OF TECHNOLOGY

The desire to avoid both technological determinist accounts of innovation as well as the desire to avoid the full indeterminacies of subjectivist theories of technological change, has led recently to a number of attempts to offer a more sociologically sophisticated account of the ways in which technologies emerge and are accepted (Bijker *et al.* 1987; Callon 1986; Law 1987). One recent attempt draws quite directly and persuasively on the recent sociological theory of Anthony Giddens and since this echoes our own recent work (especially, Silverstone 1994*a*), we will use it as the basis for offering an account of our more specific approach to the domestication of technology.

Giddens's problematic is a familiar, and an intransigent, one in sociology. It concerns the relationship of structure and agency in social life, a concern that in turn expresses in sociological terms the familiar dilemma of determination and freedom in human action. Giddens's 'solution' to this dilemma is to posit a dialectic of structure and agency which acknowledges both the duality of facilitation and constraint within structures, and, at the same time, recognizes that structures themselves are the product of social action. The third element is that of reflexivity—the capacity of participants in social life to think about and reflect upon the nature of their actions and the conditions under which they are pursued. We participate in social life as sentient beings and, in our actions, we produce and reproduce what in turn limits our actions. The particular balance of structure and agency, and the capacity of actors to affect or deny the power of social structures embodied in institutionalized norms of behaviour, the interpretive schemes we have at our disposal, as well as our power and resources within the system, will depend on the particularities of social and historical circumstance. Structures then are both objective and subjective. Actions are both determined and free. Social action involves both the constitution and communication of meaning. The dialectic of structure and agency is subsumed within a theory of structuration which insists on understanding the dynamics of a complex and, in the last analysis, indeterminate process (the unintended consequences of human action) which involves different actors differently and which will have different consequences and outcomes for the conduct of social life.

While acknowledging the large-scale implications of technological and media change for society as a whole, Giddens (1984, 1991) has not incorporated such analysis into a more detailed concern with structuration though it is clear that his approach can be made quite relevant to both. As Orlikowski notes: 'Technology is created and changed by human action, yet it is also used to accomplish social action' (Orlikowski 1992: 405). This seemingly banal observation provides the opening of an account of technology which privileges its duality and its flexibility. The duality of technology mimics precisely the more widely assumed duality of structure. It is both an objective force and a socially constructed product. As she says in relation to technology's *duality*:

[T]echnology is physically constructed by actors working in a given social context, and technology is socially constructed by actors through the different meanings they attach to it and the various features they emphasize and use. However, it is also the case that once developed and deployed, technology tends to become reified and institutionalized, losing its connection with the human agents that constructed it or gave it meaning, and it appears to be part of the objective, structural properties of the organization. (Orlikowski 1992: 406)

In relation to *flexibility* Orlikowski points to the capacity of users to interpret, appropriate, and manipulate technology in various ways depending on opportunity and context but she also points to the likelihood of this becoming increasingly limited as the routines associated with use (in organizations or in other social settings) become more institutionalized. However, she insists on the user's capacity to intervene at any time during a technology's existence, drawing on the phrase 'interpretive flexibility' (following Pinch and Bijker 1984) to overcome what she sees as the artificiality of the distinction between design and use (Orlikowski 1992: 409).[13]

While acknowledging the importance of attending to multiple levels of analysis, Orlikowski's own attempts at explanation do not satisfactorily address (as she admits) the specific properties of organizations, nor (in our view) the specific properties of software as a technology. Given that her case-study involves the creation of a software tool and its use within one and the same organization one might wish to be cautious before accepting her account as conclusive. Indeed, it is precisely at the interface of production and use, design and domestication, that the definition of technology is at its most vulnerable and at its most interesting.

In particular, it is worth pointing to the weakness of Orlikowski's understanding of the relation between design and use. Design, as we have already argued, has at least three characteristics. Each needs to be understood in its own terms as the product of the structured (and structuring) actions of all those involved and which are of different, albeit related, orders. Two points especially need to be made. The first is that it is the market, in the form of commodification, which intrudes as a crucial component in the definition of technology precisely at the point where design and use confront each other. The second is that it is the specific characteristics of distinct technologies, and technological 'genres' (such as media and information technologies) that need to be addressed. As we have pointed out radio was more than simply an object, more even than a technology. One could hardly make sense of that history without understanding its status and role as a broadcast medium.

Nevertheless Orlikowski's approach is suggestive. She offers a processual account of technological innovation which does recognize that different kinds of technologies will have different routes towards their complete definition and that there will be an indeterminacy in those routes, as users (at every stage) intervene in the structuring process. The final meaning and significance of the technology is not predetermined nor prescribed—nor perhaps ever final. It is historically and sociologically situated in quite particular ways. It is the particular, though increasingly uneven, struggle between the insistence of structure and the resistance of agency that makes that history and sociology so instructive.

THE DOMESTICATION OF MEDIA AND INFORMATION AND COMMUNICATION TECHNOLOGIES

Our concern has been with the dynamics of the complex process in which users and consumers are seen to be active, not just passive, participants. We have suggested that media and information and communication technologies are not just material or functional objects but have a powerful symbolic charge. This symbolic charge is itself the product of the activities of those who, together, design, market and use technologies. Ultimately the meanings and significance of all our media and information products (both hardware and software) depend on the participation—with varying degrees of commitment and interest—of the user, the consumer.

From these starting-points it can be suggested that the innovation process—especially as it relates to media and information and communication technologies and services—is a process of domestication. It is a process of domestication because what is involved is quite literally a taming of the wild and a cultivation of the tame. In this process new technologies and services, by definition to a significant degree unfamiliar, and therefore both exciting but possibly also threatening and perplexing, are brought (or not) under control by and on behalf of domestic users. In their ownership and in their appropriation into the culture of family or household and into the routines of everyday life, they are at the same time, cultivated. They become familiar, but they also develop and change. New technologies and services are found a place alongside or as a replacement for existing ones. As such, domestication is fundamentally a conservative process, as consumers look to incorporate new technologies into the patterns of their everyday life in such a way as to maintain both the structure of their lives and their control of that structure (Thrall 1982). Indeed, as Rogge and Jensen (1988) have argued in relation to the television and video, new technologies are defined very much in accordance with the dominant and insistently gendered character of domestic life, an insistence which is expressed in the gendering of space and time as well as in the division of labour.

In one way or another these developments and changes, the product of the work consumers do in taking possession of new technologies and services, hardware and software, feed back into the innovation process: reinforcing it, diverting it, sometimes rejecting it more or less completely. This is why the results of the innovation process are so difficult to predict. It is also worth noting that in the dynamics of domestication, and in the mostly unequal politics over the meaning and influence of new technologies, it is not only the significance of the technologies which shifts, for domestication also affects the domesticator and his or her culture. Domestication is not simply a one-way process. We are not suggesting that the rejection of a notion of technological determinism is to be replaced by an equally absurd notion that consumers of technologies, users of information services, or audiences of programmes are not affected by their participation in media and information culture. On the contrary. However, it is precisely the social, political, and economic dimensions of the struggle over meaning and influence which are at issue.

The particular character of our own domesticity is the product of, and must be understood within, the various social, cultural, and technological networks that lock a given household into the constantly shifting structures of everyday life. Our domesticity is mobile and uneven in quality, alternatively pervious or impervious to influence. It is vulnerable to changes in the technological environment, but it is also, and just as significantly, vulnerable to changes in demography or economic cycles. There is a history of our domesticity which is closely linked to the history of industrial society, a history marked perhaps above all by the shifting relations between home and work, the institutionalization of a gendered division of labour and a progressive intrusion of the State into the private affairs of the family (Donzelot 1979). In other words, our domesticity is of a piece with, but distinguishable from, the wider set of relations that constitutes our everyday life. Indeed, one of the major questions confronting an examination of the role of media and information and communication technologies in everyday life is precisely their significance in shifting, extending, transforming, or undermining the boundaries that separate our private from our public lives (Meyrowitz 1985).

Nevertheless, the increasing presence throughout the twentieth century of media and information and communication technologies in our homes and households; and their increasing centrality in the patterns of communication and the overall culture of contemporary society have made it impossible to think of our domesticity (never mind the public sphere) without them. But the reverse is also true. No account of technological innovation can ignore the particularity of that domesticity and the processes by which it is sustained.[14]

It has also been argued (Douglas and Isherwood 1979; Miller 1987) that our domestic lives are to a greater and greater extent defined by and through our consumption of the objects and meanings of contemporary capitalism. It follows that consumption can be considered not just as the lifeblood of our industrial system and our participation in a public world but also as the lifeblood of our domestic world. It is precisely in and through consumption that the two are forever intertwined.

CONSUMPTION

Pursuing this latter observation—that is, the centrality of consumption to an understanding of domestication—we have developed a model of the consumption process that throws some light on to both the general and specific dimensions of the ways media and information and communication technologies are accepted (and by extension resisted or rejected). We have suggested that the specificity of these technologies in this process is defined by their double articulation in domestic life. This notion of double articulation has been referred to already in this chapter and it needs to be clarified as it is crucial in identifying the specificity of media and information and communication technologies.

The notion of double articulation is derived from the work of André Martinet (1969) who understood the unique capacity of natural language to convey complex meanings to be the result of the articulation of both its phonemic and morphological levels. Sounds (without or with very little meaning) were a precondition for words or signs (with meaning). The meaningfulness of natural language is made possible by, and requires, both. As we argue here, the meanings of all objects and technologies are articulated through the practices and discourses of their production, marketing, and use. The technical dimensions of the machine, its design, its image constructed through advertising and other public displays, are of a piece: what is being communicated is the meaning of the commodity as object, and while this meaning is significant, it can only be a precondition for the meanings that are generated by the texts and communications of the technologies—their messages.

Media and information and communication technologies, therefore, and obviously, carry a second level of meaning in their texts and messages, a level of meaning whose communication depends on their distinct status and meaning as objects and machines, but which is far from reducible to that status. Media and information and communication technologies—perhaps pre-eminently television itself—provide in their programmes, narratives, rhetorics, genres, and software, the basis for a second articulation in culture. This second articulation and the meanings that are produced through it only become available as a result of the prior articulation of the technologies themselves.

It is in this sense that the cultural value of media and information and communication technologies lies both in their meaning as objects— embedded as they are in the public discourses of modern capitalism as well as the private discourses of home and household—and in its content, in its messages, which are similarly embedded (Morley and Silverstone 1990). The consumption of both technology and content defines the significance of these machines and services as objects of consumption. It is in this sense that we choose to think of them as doubly articulated.

This formulation offers an implicit critique of McLuhan's argument that the 'medium is the message' for what we are suggesting here requires us to confront both medium and message as meaningful, and furthermore as meaningful not in any essentialist but in a fundamentally sociological sense.[15]

Consumption consists, we suggest, of three distinct but related dimensions. The first is *commodification* in which, as a result of the activities of industrial designers, public policy-makers, regulators, and market-makers, specific claims for a function and for an identity of a new product or service are made—in this case a new machine, a new piece of software, or a new television programme. Commodification is necessarily a prior but not necessarily a determinant dimension of consumption. Through it, objects and services enter the public domain of, in Daniel Miller's (1987) terms, alienated meanings until the commodities which embody them cross the threshold into a private sphere. Commodification refers to the industrial and commercial processes that create both material and symbolic artefacts and turn them into commodities for sale in the formal market economy. It also refers to the ideological processes at work within these material and symbolic artefacts, work which defines them as the products and, in varying degrees, the expressions of, the dominant values and ideas of the societies that produce them.

Commodification necessarily depends on a dimension of imaginative work that potential or actual consumers undertake as they participate, willy-nilly, in the consumption process. The work of imagination is, however, contradictory. Commodities are constructed as objects of desire within an advertising and market system that depends for its effectiveness on the elaboration of a rhetoric of metaphor and myth: a seduction of and through the image (Ewen and Ewen 1982; Ewen 1984; Leiss *et al.* 1990; Williams 1980). Yet such imaginative work (and work of the imaginary), both that of the advertiser and the consumer, is a frustrating experience. It is frustrating because of the limits imposed by consumption itself. For every act of successful consumption, as Alfred Gell (1986) suggests, there are many failures, failures defined by economic constraints, inadequate resources, and limited objects and products. However, even 'successful' consumption involves failure, for such failure is both endemic and necessary if consumption is to continue. Jean Baudrillard (1988) describes consumption as a kind of general hysteria, based upon a fundamentally insatiable desire for objects, a desire that can never be satisfied. Needs, therefore, can be neither defined nor fulfilled since consumption is based, not on a desire for objects to fulfil specific functions, but on a desire for difference, a desire for social meaning (Baudrillard 1988: 45).

As other commentators (for example, Schwach 1992) have noted, in the actual practice of consumption, goods are imagined before they are purchased, prior to any loss of illusion that comes with ownership. Purchase in this sense is, potentially at least, a transformative activity, marking a boundary between fantasy and reality, and opening up a space for the consumer-user's imaginative and practical work on the meaning of the technology or text, either as a compensation for disappointed desire or as a momentary celebration of its fulfilment.

The second dimension of consumption is *appropriation* in which socially located individuals (individuals distinguished by class, age, gender, ethnicity, and as members of families or households) accept enough of the relevance of the publicly defined meaning of something to their own circumstances to buy and then accept the new object or product into their own domestic environment. This process of appropriation is an active and, to a greater or lesser degree, also a transcendent or transformative one. For an object or technology to be accepted, as we have already suggested, it has to be made to fit into a pre-existing culture. It has to be found a space, literally, in the home. We call this aspect of appropriation *objectification*. Such objectification, like so much of consumption, is fundamentally reflexive, since it is possible to suggest that material and symbolic artefacts of all kinds (including in the context of this argument both the machines and the messages), in their physical and discursive arrangement and display, provide an objectification of the values of those who feel comfortable, or identify, with them. So an understanding of the dynamics of objectification of commodities within the household will also throw into strong relief the pattern of spatial and symbolic differentiation (male/female, adult/child, shared/ contested, public/private) which provides the basis for a given domestic geography. And of course vice versa: an understanding of domestic geography, both in its physical and symbolic aspects, will provide a basis for understanding the specific significance of objects played within it.

Media and information and communication technologies also have to have a function. They have to be fitted into a pattern of domestic use in domestic time. We called this aspect of the work of appropriation *incorporation*. Two points need to be made. The first is that the functions may be somewhat removed from those intended by designers or advertisers. Functions may, for example, change or disappear, as in the case of those home computers originally bought for educational purposes which have become games machines or have been relegated to the tops of cupboards or the backs of wardrobes. The second point is that incorporation may release time for other activities. It may facilitate control of time, for example, in the time shift capabilities of the video or the microwave. Or it may simply enable some times to be better spent, for example in the use of the radio as a companion for the tea-break, or as part of the routine of getting up in the morning.

Such appropriation is complicated, of course, by the social dynamics and politics of families and households. There are conflicts over use and location; over ownership and control. There are anxieties to be dealt with: about the disruption a new product might introduce into the security of familiar routines and rituals, the challenges it might create to an individual's competence or skill, or the threats it offers to a household's moral values. The pressures to accept or reject, as well as to modify, the meaning of a new object or technology are not generated only by the complex politics of domestic life, but come too from the conflicts between domestic and public values; from, for example, the competing claims of parents or peer groups.

The third dimension of consumption is one which reconnects the household into the public world of shared meanings and the claims and counterclaims of status and belonging. We call this *conversion* and it signals the importance of the need to legitimate one's participation in consumer culture in the display of competence, and ownership. Such conversion extends from showing off the new video-phone to the neighbours to the endless gossip about the latest episode of a soap opera. It is also the case that it is through conversion that the spiral of consumption continues to turn, for in our converting activities (and not just through initial rejection) those involved in commodification (producers, regulators, advertisers, and the rest) learn about consumption and may or may not alter their products and services to fit what they think they have learned.[16]

These three dimensions of the process of consumption are also three moments in both the domestication of new technologies and services, and of the construction of the domestic itself. Media and information and communication technologies are central because they are themselves both objects to be consumed and the facilitators, through their status as media, of consumption. Through our involvement with them we learn how to consume and what to consume. And through our involvement in consumption, we learn to display who and what we are. In this lies media and information and communication technologies' distinctively reflexive role in everyday life.

Our understanding of the place and significance of information and communication technologies in the domestic sphere and the process of domestication which puts them there is therefore dependent on exploring some of the key social and cultural (as well as economic) dimensions of that domesticity. This begins with a concern with the obvious but crucial differences that differential access to economic resources create. This is a matter not just of the level of disposable income or economic capital, it is also a matter of the allocative rules of family or household income: the pattern of its earning, its management, and its control (Pahl 1989).

THE DOMESTIC CAREERS OF INFORMATION AND COMMUNICATION TECHNOLOGIES

Digging deeper into the ways in which information and communication technologies are domesticated involves consideration not just of the fine grain of family and household life but also of the often contradictory dynamics of the process of domestication itself, as new technologies and services offer themselves to the contemporary household. Central to an understanding of these processes, must be an awareness of the complex politics of family life: of the relationships between partners as well as between parents and children, of the politics of control and ownership of space, time, and technologies, of the persistence of the gendered divisions of labour in the home. It must also involve awareness of the tensions and conflicts that are the product of change as the family moves through its life-cycle, and as it is buffeted by the squalls of everyday life. Of course it is not just individual families which change. Central also to an understanding of the dynamics of the domestication process are much more fundamental changes in demography and in the political economy of the household: in the increase in lone-parent households, elderly households, in divorce and unemployment rates, and in the changing relationships between home and work consequent upon both industrial restructuring and the opportunities offered by new technologies.

Equally central must be an awareness of technological change itself. This is quite clearly taking a number of forms. Once upon a time, the distinct technologies that define the domestic field—the telephone, television, radio, and the computer—offered relatively discrete functions and discrete experiences, respectively communication, reception, and information processing. These functions and experiences are becoming less distinct and more integrated and interrelated. Indeed, the implications of this convergence are far-reaching—at least potentially—not only in the shift of attention and marketing away from the technologies themselves to the software, programs, and services that they deliver, but also in their integration into networks, global as well as local, that involve their users in a teletechnological system (Silverstone 1994a) that is both ideological—it carries with it a set of values and constraints that are barely visible but consistently intrusive, and increasingly interactive.[17] Less certain of course is how far, how fast, and in what ways, this demonstrable technologically derived convergence will be appropriated into the household. Indeed, to a degree, a kind of social and cultural convergence has preceded that offered by the technology. Families, households, and individual members of both are conventionally and habitually quite adept at a kind of seamless shifting from one technological input and resource to another as well as being adept at their simultaneous use. Households, from this perspective, have long been convergent in their own technological culture, a convergence that is, as we have already noted, both conservative—new technologies and services have to be adapted in their adoption—and innovative—mostly in the sense that they are open to new and better ways of doing familiar things.

The domestication of information and communication technologies and services is therefore as complex and contradictory a process as design. It involves practical and symbolic work. It depends on the particular balances of power between key players, only in this case the players are in the consumption rather than the production game. It depends on available resources. And it depends on the particular circumstances of decision-making as well as their preconditions in the culture of the family and household. As a consequence, domestication is a more or less continuous process in which technologies and services are consumed—socially and culturally both chewed and swallowed—and, through the process of consumption, are given meaning and significance.

Igor Kopytoff (1986; and see Haddon and Silverstone 1994) has offered an analysis of what he has called the biography of technology, by which he means the changing identity of technologies and commodities as they pass through their own life-cycle. He suggests that the tracing of the biography of, for example, a car in Africa as it moves from the factory to the scrap heap, would simultaneously reveal the nature of the changing social and cultural as well as political and economic relations that are involved in its progress and decline. It would provide a route into the comparative analysis of social shaping; for the same car in France or Russia would by virtue of its distinctiveness reveal much about the dynamics of French or Russian as opposed to African culture.

This is extremely suggestive. Information and communication technologies have just such biographies, and it is in their biographies that a number of different temporalities are condensed. The history of the technology as such (of television as a machine or a broadcast form, for example; or the home computer) coincides in consumption with the history of its purchase and use within a given household. Tracing the biography of technologies is the task in which we are engaged, though with a number of qualifications and extrapolations. First, we note that a technology's biography is crucially one that involves both production *and* consumption, design *and* domestication. Second, we reiterate that such a biography, what we have called a technological career to signal both its functional significance and its dynamic, is the product, and is expressive of a series of interweaving temporalities. These include the history of the technology itself, the history of the specific example of the technology, the history of the family and the biography of the individual, the dynamics of the family's life-cycle, and the particular orientation in time that the family adopts (Silverstone 1994*b*). Third, we resist the idea that all biographies are linear, progressive, or conflict and contradiction-free.

We can trace such biographies and careers in the spaces and times of the household and we can follow their implications both for the status of the technologies concerned, and for what their movement through social and cultural space tells us about the conditions and consequences of their domestication. These processes are particularly visible in teleworking households in which technologically facilitated new forms of work more or less dramatically shift the technological culture of the household as a whole. By contrast, in the households of lone parents where many information and communication technologies come to their final resting place, the absence of anything beyond the basic and the old is ample testament of the need to take into account the huge variation in our capacity to appropriate the new information culture (Haddon and Silverstone: 1993, 1994).

The arrival of a new technology in a household is a social and cultural as well as a technological event. So, too, is the redefinition of the use of a technology, or its enhancement with the addition of new peripherals or new services. These various dimensions are fundamentally intertwined. Fundamental is the challenge such novelty poses to the established routines and rituals of family life. A teleworker, for example, will need the telephone for his or her business. A second dedicated line may well be the solution to the problem of managing social calls or indeed releasing the phone from the blocking of the omnivorous adolescent in the family. But if resources do not permit, and even if they do, the family culture around the use of the telephone and the social skills required for its answering will be significantly affected as the rest of the household has to learn to be able either to distinguish between lines, to learn new answering and message taking techniques, or to abstain altogether from making their own calls at certain times of the day. Equally, the arrival of a new television will tend to relegate the old one to a secondary status. No longer will it be in the control of the dominant figure of the family or be perceived to be the family's to share. The discarded technology will be denied its place at the power point of the household. It will shift physically to less significant spaces, to the kitchen (where old black-and-white television sets tend to languish) or to a child's bedroom (usually the boy's) where it will become part, especially as the child gets older, of a converging teenage techno-culture of sound, vision, and computer games. It might also shift out of the household altogether as a gift to a disadvantaged friend or relative. Many television sets and video-recorders, not to say telephone handsets and obsolescent games-playing consoles, continue their careers in this way. In their progress, they articulate the social and cultural divisions in contemporary society and the differential access to our increasingly pervasive information and communication-based culture.

We can identify a number of dimensions of this process of domestication in which both the career of the technology and the culture of the household are involved. As the technologies shift around the household so, too, do the skills and competencies associated with their use, though not necessarily in a uniform or uncontested manner. This process of seepage may enable children to become incorporated into an increasingly sophisticated computer-based family culture, or vice versa, peer-group encouraged and supported children's involvement in games or computer activities may attract their parents. That incorporation, however, might equally be contained or resisted by gender or age-based divisions in the culture of the household in which the women or parents are disqualified or disabled when it comes to programming the video-recorder or confronting the keyboard (see Turkle 1986, 1988). This technological seepage is, of course, social. It involves not just the physical movement of machines around the household, but their changing functions as the households and the technologies change. The status of a new machine will be affected not just by its replacement by the latest version or the addition of new functions or services, but by the changing demands on it from different members of the household. Machines are inherited, technologies are replaced and displaced, and patterns of use are reconfigured. All this involves both regulation and reregulation as well as the management of conflict over access and use.

It involves, above all, the politics of the household. Access to media and information technologies is, in different ways, empowering. To be able to take control of the television remote control, or to be the one member of the family who can effectively programme the video-cassette recorder; to be able to limit access to a new computer by insisting that it be sited in one part of the house rather than another, all of these dimensions of the relationship to technology are expressive of the dominant age and gender-based politics of the family as well as reinforcive of them. Access and control, as well as denial of access and lack of control, are fundamental to an understanding of the use of communication and information technologies in the domestic, as in other, spheres and they are also fundamental to their meaning.

Equally, access to information and communication technologies also has consequences for a household's participation in the wider activities of everyday life. Household boundaries are extended through gendered social and familial networking on the telephone, on the one hand; and by television's electronic reach, its capacity to bring global information and images into the front room, on the other. It is also extended within the neighbourhood or community as ownership of new machines, faxes, or photocopiers, as well as the expertise associated with computing, come to be seen as a shareable and accessible resource. It is extended even further by the increasing mobilization and personalization of communication and information technologies, as walkmen and mobile phones offer a new kind of nomadic access and media participation, constant availability and increasing dispersal of information consumption.

Our argument is intended to draw attention to the duality of technology both within the domestic sphere as well as at the design–domestication interface. This duality is neither singular nor unchanging. The double lives of technology as well as the double articulations of communication and information technologies in the home offer a complex matrix of opportunities and constraints. Information and communication technologies are both shaping and shaped in this culture and through the social dynamics of domestication. Some are rejected out of hand, while, for others, the process of domestication is one of marginalization and gradual discard. Others become increasingly incorporated into the infrastructure of the household, creating their own dependencies and offering their own contribution to the ontological security (Silverstone 1994*a*) of the household. The freedoms that we have to impose our meanings on these machines, and to impose significance on the services and software that they offer, depend on the amounts and character of what Pierre Bourdieu (1984) would call our economic and cultural capital. However, they also depend on the political and cultural dynamics within the household and in the household's economic relationship to the wider world.

These freedoms are not infinite. They are constrained as much by the rhetorics of the technology, expressed through design and marketing, as by the particularities of domestic culture. Both are constrained by the presence of an established public culture of technology and by the existing systems and networks which both enable innovation and guide its progress. Orlikowski makes the point that technical systems quickly become reified, and that the opportunities for creative consumption are short-lived. This is as true for households as it is for organizations. This conservatism is at the heart of the innovation process, and our domestic lives are grounded in, and depend upon, a high degree of 'taken for granted' activity. Domestication involves the drawing of technology's sting but its results are not always easily predictable. Information and communication technologies may offer evolutionary rather than revolutionary possibilities (see Eisenstein 1979; Forester 1989: 19–96; Marvin 1988), but households themselves change, both internally and in their relationships to the external world and that change too has consequences for the innovation process as a whole.

CONCLUSION

Very little work either in the field of science and technology studies or in media studies has addressed the innovation process as a whole. Yet, it is in the dynamics of this process that we can avoid the reifications and false certainties of partial and unsociological accounts. The flaws in many of these are the result of attempts at explanation that grant technology an unmediated sway over social life. These accounts fail to recognize the specific characteristics of communication and information technologies, adopt a singular and abstracted notion of rationality and efficiency as measures of consumption and use, and misread the dialectic at the heart of technological change. In this chapter we have offered a more inclusive account. In the juxtaposition and conjunction of design and domestication we have identified and elaborated upon the forces and uncertainties of innovation as they relate to the household and to the acceptance of new communication and information technologies in everyday life. At a structural level, these arguments have a wider relevance, for design is a universal feature of the innovation process and domestication takes place in all social environments confronted by the challenges of new technologies and systems. This is particularly true at the level of the organization where similar institutional patterns and forces conspire to divert and undermine the best intentions of systems and hardware designers. Indeed, as Paul Quintas argues in Chapter 3, software design is increasingly a matter of embodying the culture of the user organization into the fabric of the technology.

The household provides a clearly identifiable case of a situated reality in which the norms of economic and social behaviour are defined, not by abstract principles, but by the particularities of private and personal values. The domestic economy is a moral economy in the terms in which E. P. Thompson (1971) describes the traditional economic behaviour of rural England in the late eighteenth century confronted as it was by an expansive, unfamiliar, threatening, and unfeeling market. It is a moral economy also in the sense in which anthropologists (Cheal 1988; Parry and Bloch 1989) have used the term, drawing attention to the transcendent capacity of families and households (within their private sphere) to determine the character and direction of their domestic lives. What they identify is something more akin to a gift economy in which the long-term interests of the participants take precedence over the short-term and instrumental relationships that mark economic behaviour in the wider environment. There are clear links here between this perspective and the tradition of institutional economics discussed by Robin Mansell as they relate to wider institutional patterns of behaviour and values in organizational settings. In both, there is a stress on the historical and social specificity of economic behaviour which requires recognition of indeterminacy, uncertainty, and change, as well as that innovation in a contested process. The politics, economics, and sociology of innovation is always skewed and uneven, fought out at the interface of design and domestication. These relationships are themselves complicated by the competing interests within producing and consuming groups as well as those between them. It can be objected that framing the design/domestication interface as contested misreads the essential facilitating and reinforcing relationships between production and consumption: the necessary symbiosis of all market economies. On the contrary, we argue that this relationship must be seen as a dialectical one whose outcome cannot simply be read off from intention or precedent. Technical and institutional innovation, similarly, are processes that must be understood in their essential tensions.

NOTES

1. The study of information and communication technologies has been significantly enhanced by perspectives derived especially from media studies, where the status of media as technology has been (with the exception of the much criticized McLuhanite tradition) relatively underdeveloped. More germane is the recent concern with audiences and the reception of, especially, television programming increasingly contextualized methodologically within an ethnographic approach, and increasingly theorized from the point of view of viewers as active participants in the media/consumption process (for recent reviews see, Morley 1992; Moores 1990; Silverstone 1994a).

2. The research conducted between 1987 and 1990, and 1991 and 1994 was funded by the Economic and Social Research Council in the United Kingdom under its Programme on Information and Communication Technologies. For discussions of the first phase of the research, see Silverstone (1994a); and chapters in Silverstone and Hirsch (1992); for reports on separate aspects of the second phase research, see Haddon and Silverstone 1993, 1994, 1995).

3. Clearly, this definition is an emergent and unstable property of the innovation process or less clearly defined depending on the specific technology or service being marketed and the point it has reached in its life-cycle as a commodity. It can be argued, for example that the home computer took some time to emerge clearly as a functional object (Skinner 1992).

4. 'Media are not fixed natural objects; they have no natural edges. They are constructed complexes of habits, beliefs and procedures embedded in elaborate cultural codes of communication. The history of media is never more or less than the history of their uses, which always lead us away from them to the social practices and conflicts they illuminate' (Marvin 1988: 8); and of course in relation to the last point, also the reverse.

5. Broadcasting itself was *designed* and as many have argued (esp. Williams 1974; and Scannell and Cardiff 1991) the particular character of radio as a technology was not determined by anything in the machine but was the

product of a more or less motivated attempt by State, manufacturers, and broadcasters to control the new technology, at least in the UK and Europe, for 'the public good', but above all as a centralized broadcasting technology. The fact that radio was technically quite able to offer interactive communication (and was originally designed to do so) is often overlooked.

6. See once again, Marvin (1988: 76) writing on the introduction of electricity into late nineteenth-century homes: 'Home was the protected place, carefully shielded from the world and its dangerous influences. New communications technologies were suspect precisely to the extent that they lessened the family's control over what was admitted within its walls. Householders resisted both the symbolism of outside intrusion and its physical expression in wiring, a tangible violation of intact domesticity.'

7. 'There are those for whom TV is an object, there are those for whom it is a cultural exercise: on this radical opposition a cultural class privilege is established that is registered in an essential privilege' (Baudrillard 1981: 57). Baudrillard is pointing rather crudely and inadequately to this double articulation in which television (and other communication and information technologies) are doubly inscribed into the domestic, both as (social) object and as (cultural) medium (see Silverstone 1994a: 112–14).

8. de Sola Pool (1977); and Noble (1986 cited in Keen 1987: 9): 'close inspection of technological development reveals that technology leads a double life, one which conforms to the intentions of designers and interests of power and another which contradicts them—proceeding behind the backs of their architects to yield unintended consequences and unanticipated possibilities.'

9. See Dorothy Smith (1978) whose analysis of reports of mental illness has been extremely influential.

10. The danger in this form of analysis is that it tends to be structured through a perception of the industrial world as being (in the end) manageable and relatively conflict free. This may in a given situation (and for a given time) be the case, but it should not be taken for granted.

11. The case-study was generated as a result of a series of interviews with producers, advertisers, retailers, trade journalists, and early users, together with supporting documentary analysis. It was very much an exploratory study. The research offered an account of the emergence of a new technology at a precise historical point, indeed at a point at which the future of the product and the speed and character of the acceptance of what multimedia in general was offering was still very much uncertain.

12. At the time of the study PC-CD ROM multimedia systems had not yet made a significant impression on the domestic multimedia market. Their subsequent success is instructive.

13. Orlikowski's empirical example concerns the creation and use of a series of design tools in an international information technology systems consulting company which she shows have organizational consequences as well as being negotiable in use. She makes the point that the increasing power of the organization through time (and its own 'understanding' of what the technology can do, as well as its perceived functions and benefits) increasingly limits the capacity of users and clients to challenge the taken-for-granted integrity of the technology so that the structures of both

organization and technology progressively coincide. The point of course is that it is in this process of mutual adjustment that both technologies and organizations change, and that the one does not determine the other—in either direction.

14. It has been argued that this intense intertwining of the social and the technological in the home has created a new form of hybridization in which technological and social processes are indistinguishable (Jouet 1993).

15. As Rohan Samarajiva has helpfully pointed out it is possible to distinguish between three, rather than two, dimensions of message and articulation: that grounded in the object, that in the symbolic environment to which the technology gives access, and that of specific 'programme' messages. This is correct. Our argument in the body of the chapter, however, is based on the privileging of the first two. The last—the single text/message—is clearly present and, on occasion, of importance but, as many analysts have noted, it is itself dependent on the generic symbolic environment and insignificant in terms of influence when compared with the overall flow of media messages (see Gerbner 1986).

16. For a more extensive and more thoroughly contextualized account of this model of consumption, see Silverstone (1994a: 104–31).

17. We can begin to recognize both the values and their force in the perverse mirror of the internet where, at least for the time being, they are reflected in their absence.

3

Software by Design

PAUL QUINTAS

> The bad news is that, in our opinion, we will never find the philosopher's stone. We will never find a process that allows us to design software in a perfectly rational way. The good news is that we can fake it.
>
> (Parnas and Clements 1986: 251)

INTRODUCTION

The previous chapter has explored the concepts of design-in-use, suggesting that technological design and innovation are processes completed in the consumption of information and communication technology products within the domestic environment. Conventional notions of the spatial, temporal, institutional, and social location of design activity have thus come under critical scrutiny. In this chapter we continue this critical treatment, moving away from the domestic environment to investigate the activities of 'production'. Our focus is on software and, in particular, the processes by which software systems are designed, the agencies involved, and the organizational contexts in which design takes place. Sharp divisions between 'production' and 'use' are not always appropriate in the software arena and the heterogeneous and temporally changing relationships between software developers and users are the central focus of this chapter.

Why look at software? First, we need to look at software because it is economically and socially significant. It is pervading nearly all aspects of our lives. In the 1990s, software ranks among the most important technologies for the world economy (in fact, some suggest it is the most important). OECD data show that in 1991 software expenditure exceeded information technology hardware expenditure for the first time in history (OECD 1994b).[1] World software trade grew at over nine per cent between 1991 and 1992 while hardware trade fell by three per cent. Significantly, the $US 184 billion recorded world software trade in 1992 does not include software developed for in-house consumption. In-house activity does not appear in trade statistics and the OECD estimates that its inclusion would double this figure to around $US 370 billion.

In spite of its importance, software and the processes by which it is designed and developed, are poorly understood by policy-makers, economists, and scholars of technical change. Current understandings of technological innovation and diffusion are dominated by hardware models based on traditional industries that have only limited relevance to software. The analysis in this chapter is based on the premiss that software is sufficiently different from other technological phenomena as to require a reconceptualization of the way we think about the design and use of technology.

In the following section, we discuss what software is in some detail. Software refers to the programs or instructions that control and operate computer and telecommunication systems or provide the applications or functions that run on these systems. As well as the familiar software applications such as word processors and spreadsheets, and the databases and information systems that support the activities of most organizations, many programs are also embedded in systems and manufactured products such as electrical goods, vehicles, and machinery. Increasingly, the technologies with which we interact at home, in the high street, or the workplace, or, indeed, in the workplace at home, are software machines. Even the product characteristics that we experience in our dealings with apparently 'hard' technologies such as washing-machines, microwave ovens, or cars, are now becoming software creations. This is not just because computer-aided design software is used in the design of such products, which it may well be, or that software provides the user interfaces through which we interact with the hardware. It is because software embedded in these products controls machine functions and, therefore, the characteristics of the technical system.

The theme of redesign and innovation *in use* that was the focus in Chapter 2 takes on added dimensions when we consider that the software that controls machines and systems may be reprogrammable, thereby enabling changes to be made to the characteristics and functions of the machine. In addition, software systems are being installed that enable machines to learn their user's patterns of use and automatically optimize and tailor the system's performance to that individual user.[2]

Software is, as its name suggests, highly malleable. The development and diffusion of information and communication technologies is inexorably linked to software, not only because hardware is useless without it, but because software enables hardware to be reprogrammed to meet a wide range of requirements. This allows standard hardware components, especially semiconductors, and products, for example, personal computers or car engine management systems, to be produced in high volumes, leading to reduced hardware costs while retaining very high levels of flexibility through the malleability and tailorability of software. It also means that many of the key design issues for information and communication technology systems and products are addressed within the software development process. These key design issues include the extent to which the software will enable and support user participation in its own evolution and change over time.

A central theme in this book is the analysis of the problematic relationship between design, use, and the dynamically changing capabilities of human agents. This chapter is especially concerned with the relationship between developers and users of software. We show that the identification of software developers is itself a problematic issue. There is no universal model that we can adopt to illustrate the general picture of the developers' relationship with the users of the programs they develop. Developer–user relationships change significantly over time, and particularly throughout the careers or life-cycles of software 'products'.

In looking at the software development process we also need to consider the design of software work; that is, how developers' work is organized and controlled. This gives clues to the nature of software as a design- and labour-intensive creation as well as providing insights into the changing relationships between professional developers, management, and software users.[3]

Consideration of software and its development in these terms presents a challenge. The central questions addressed here are: who develops software; when and where do software design occur; who are the users and when are they involved in the design process; what is the relationship between developers and users, and how does this relationship change over time; how do design decisions made at different points in time enable or constrain future choices; and how do patterns of innovation in software development and changes in acquisition strategies affect the user–developer relationship? In addressing these questions we examine a number of contradictions, such as the tension between software's inherent flexibility, on the one hand, and, on the other, the ways in which initial design decisions and sunk investment costs prefigure use patterns and constrain the degrees of freedom for future developments.

WHAT IS SOFTWARE?

The separation of software from hardware, enabling hardware to be reprogrammed to meet a range of different tasks, is crucially important in any explanation of the current importance of information and communication technologies and their widespread diffusion. It is the fact that they are *re*programmable that marks computer systems as being distinct from calculating machines. This idea pre-dates semiconductors and electronics. The original concept of a general purpose reprogramm- able computing machine is attributed to the Cambridge mathematician, Charles Babbage (1791–1871), who, between 1833 and 1837, designed a general purpose 'Analytical Engine'. Here the program was stored on punch-cards, an idea Babbage borrowed from Jacquard's early nineteenth- century weaving looms. Babbage wrote:

The Jacquard loom weaves any design which the imagination of man can conceive. The patterns designed by artists are punched by a special machine in sets of pasteboard cards and when these cards are placed within the loom, it will weave the desired pattern. (Rabel 1948: 115)

This quotation provides a useful example of the role of the program within the hardware system. The conceptual design is represented in the program which is translated into a sequence of holes in the cards. This example also relates to the questions being addressed in this chapter: who is the 'user' of the Jacquard system, what happens to the skills of the weaver, and can changes to the program be made by the loom operator?[4]

And so software, at this simplest level, provides the control functions for hardware. Today, the term, software, covers the programs, sequences of instructions and procedures that store, organize, and manipulate data in information and communication technology systems.[5] Software programs are written in logically based languages that can be processed by computers. The French word for software— logiciel—conveys this rather better than the English equivalent. In a revealing piece of jargon, the lists of symbols that form programs are referred to by practitioners as 'code'. In some ways the writing of software programs resembles musical composition or literary author- ship: symbol and language are used to construct programs which can be written down, and, indeed, in most countries software is regarded in intellectual property law as being equivalent to literary works, an issue which is taken up in greater detail in Chapter 7. On one level, software is 'ephemeral thought stuff' in the same way that the content of any language—be it mathematics, English, or musical notation—may exist only in the mind of the author. Certainly, it may be written down on paper, but it is useful only when entered into the memory of a compatible computer system.

What distinguishes software from the written word, or the scientific formula, is that it is not intended solely to be read and understood by others who comprehend the particular language. When fed into a computer system, software may enact and control physical processes such as the composition, temperature, and flow of steel in a mill, or the rate of climb in an aircraft.

Central to our understanding of software is the concept of the *virtual machine*. Software consists of sequences of notation or language, but when installed in hardware, it creates a functional system with which we can interact or communicate and which we experience as a tool or machine: a virtual machine. Software is information, instruction, and data that, when combined with the appropriate hardware, provides machine-like functions. The word-processing software with which this text is written creates a 'virtual machine' that I experience as a text editor in the desk-top computer—a virtual machine that has all the functional attributes of a typewriter plus many more. Change the software, and the virtual machine with which I interact becomes a flight simulator or a drawing-tool. Without the software, none of these machines exists— hence their virtual nature.

When speaking about software and the processes by which it is created, it is common for economists, policy-makers, and journalists to fall into the trap of assuming that the concepts and language we use to analyse and discuss manufactured products and manufacturing processes are directly applicable. However, these may be less than useful, even as analogies. For example, the concept of *manufacture* is of little relevance, since it generally is concerned with the *reproduction* in quantity of a designed artefact. The replication of any finished piece of software is incomparably trivial compared with its design and creation, as attested to by the millions of illegal copies of software packages diffusing around the globe. Similarly, software is written by authors, and we do not speak of writers manufacturing their articles and books, although we may usefully describe the printing and binding process as book production or manufacture.

The word, *development*, usefully describes the whole process of creating new software, which includes what many developers refer to as the activities of designing, building, and testing software. Software development is essentially a multilayered design process rather than a manufacturing or production process. In contrast, conventional hard-ware products have a design phase followed by production engineering and tooling-up for manufacture, so that the product may be reproduced in quantity.

Other analogies may be more appropriate. Brooks (1986: 26), an influential software practitioner and writer, says, 'I still remember the jolt I felt in 1958 when I first heard a friend talk about *building* a program, as opposed to *writing* one. In a flash he broadened my whole view of the software process. The metaphor shift was powerful, and accurate.' He goes on to say that the *build* metaphor has outlived its usefulness; since software systems have become so complex that they cannot be fully specified and designed in advance, it is more appropriate to think of *growing* software incrementally. However, growing does not imply a lack of design. Top-down design in the case of overall system design is crucial in order to provide the structure within which the software system may grow incrementally (Brooks 1986).

Software also has unconventional patterns of diffusion. Unlike manufactured products, the cost of reproduction of a software program is close to zero in comparison to the cost of its design and construction. One estimate by package vendor, Lotus, suggested that 130,000 of the 140,000 copies of their Lotus 1-2-3 product in India were illegal copies. Software, like other information products, can be shared and copied without loss and at very low cost, apart, that is, from the cost of assimilation of the software into an existing knowledge base. In addition to diffusion via illegal copying of disks, the electronic transmission of software across borders, or indeed its electronic theft, is generally undetectable. Given these characteristics, we have only the crudest measures for assessing the extent to which software products and systems have diffused throughout the world.

WHO DESIGNS SOFTWARE?

Because software development is a highly accessible technical process that requires low capital outlay and few physical resources, the barriers to entry are much lower than for conventional technical developments. In principle, anyone with a personal computer can develop programs. At the national level, the capacity to develop software is distributed throughout the economy, and more software professionals work for user organizations—that is, organizations that use information and communication technologies but do not generate revenue from sales of information and communication technology products or services—than for companies in the conventionally defined information and communication technology sector. User organizations include banks, insurance companies, utilities, manufacturing companies, and government agencies.

Pinning down the location of software designers and developers is less than straightforward, and quantitative accounts of software personnel are subject to large margins of error. Certainly, many software professionals work in software and systems companies in the information and communication technology supply industry, but, ironically, these designers and developers are in the minority. The capability to develop software is one of the most highly decentralized and distributed technological competences of the late twentieth century. Table 3.1 provides estimates of the numbers of software professionals employed in different sectors of the economy in the UK, the USA, and Europe. Although these estimates are approximate, the broad picture is

TABLE 3.1. Estimated numbers of software professionals

	UK (000s)	UK %	US (000s)	US %	Europe (000s)	Europe %
Software houses, services firms, and consultants	50	17	127	6	200	11
Other ICT firms: hardware firms, systems integrators, and telecom suppliers	70	23	423	22	200	11
User organizations	180	60	1403	72	1500	78
TOTAL	300	100	1953	100	1900	100

Source: Compiled from UK data, Oakley (1990); US data, Jones (1993); European data, Commission of the European Communities (1989).

consistent, showing that around two-thirds of software professionals work in user organizations. Capers Jones's (Jones 1993) heroic attempt to quantify this capability for all industrial sectors in the USA claims, for example, that American insurance companies employed 195,000 software professionals in 1991, with banking and finance firms employing 167,000, and aerospace manufacturers, 115,000.

This distributed capability to design and develop software suggests that conventional assumptions about the relationship between suppliers and consumers regarding technical innovation must be regarded critically in the information and communication technologies area. For example, Pavitt's (1984) widely quoted taxonomy of technical change identifies four sectoral types of innovating firms: science-based, specialized suppliers, scale-intensive, and supplier-dominated.[6] For Pavitt, service sector organizations are supplier-dominated. However, when we look at software and information and communication technology systems development more broadly, it is not the case that service sector organizations historically have been supplier-dominated firms which 'make only minor contribution to their process or product technology' (Pavitt 1984: 356). Indeed, service sector companies represent major concentrations of software design and development capability.

Software development is highly dependent on the skills of analysts and programmers. It remains a labour-intensive activity that largely has resisted automation. Demand for information and communication technology systems has, until the economic recession of the 1990s, outstripped the capacity of software developers to meet it, and software development has created large numbers of jobs, estimated at over 12 million software professionals world-wide (Jones 1993). By software professionals, we mean people who are employed in the capacity of programmer or analyst, etc. This does not mean that these individuals necessarily have a computer science or software engineering qualification. Indeed, the period of rapid information and communication technology growth and skills shortages, lasting from the 1960s until the recession of the late 1980s, meant that organizations seeking to recruit software staff brought in people with a very wide range of academic and career backgrounds. Non-technical qualifications such as languages were thought to provide a good basis from which to acquire programming skills.

In addition to those employed as software professionals, much software is developed by casual developers or amateurs. While much of this activity is aimed at producing software for the individual's own use, a large amount is used by others or becomes incorporated in products and systems. This phenomenon leads to criticism from the professional software community who argue that established branches of engineering do not suffer from the activities of casual designers of steel structures or aerospace engines (Jackson 1994). Amateur programming is bemoaned by Jackson and other professionals, for example, Macro and Buxton (1987) because, they argue, such developers do not employ professional standards of technical practice. Software project or system failures that result from 'bad practice' contribute to the lowering of the reputation of software developers in general. This reflects a fundamental dichotomy in thinking about the software development process: whether it should be the preserve of trained and accredited professional software engineers, or whether technological developments should enable non-professionals and even end-users to design and build their own systems. Both views have their advocates and are manifested in divergent paths of technical innovation and the emergence of new capabilities in the software developer communities (Quintas 1994a).

SOFTWARE DESIGN

Given the unusual and heterogeneous nature of software activity, it is no surprise that the software design process varies greatly according to the type of software, its mode of development and supply, the agencies and actors involved in the development process, and the environment in which it is to be used. Software to control the cycles of a washing-machine or car engine management system differs considerably from the software used by bank clerks to input debits and deposits, or from word-processing software, or telephone exchange switches.

A distinction should be drawn between software systems that are custom-built for a specific client or user, and those that are intended to be sold as standard products or packages that are widely available. Clearly, the relationship between designers and users is very different where the system is being tailor-made for the users, as compared to one's relationship with the designers of Microsoft Word. In between these extremes lies software that is available as a standard product but addresses a specific market niche, for example, software to manage corporate treasury funds, or dentists' practices, and software that is available as a generic product that is capable of being modified or customized to meet the specific requirements of the user, as in the case, for example, of a database management system.

Much of the professional and academic literature focuses on custom-built software development in large firms, including both information and communication technology sector firms and large user organizations. Best-selling textbooks on software development, for example, Sommerville (1992), do not address the issue of package development at all. With a few notable exceptions (Carmel 1993, 1994; Carmel and Becker 1995; Lammers 1986), relatively little is written about the design and development practices of the small software product companies, or indeed, the currently dominant product vendors such as Microsoft, Oracle, and Novell.

Software creation is generally the product of individuals working within teams. The larger the software system, the greater the number of people likely to be involved. At the extreme, the development of a new telecommunication switch may involve thousands of developers working in several continents. The task is broken down into modules and subcomponents on which small teams may work. The design of the interfaces between components, and systems testing after the integration of modules, thus become crucial factors in systems development.

Writers on software development generally distinguish between the various processes which make up the phases of design—for example, functional (high-level) system design, logical design, physical system design, module or object design, program design, and so on. In some accounts these activities are assumed to be discrete functions undertaken sequentially, but it is now generally agreed that there are iterative loops between them. As emphasized by Sommerville (1992), design is an iterative process in which the finished design emerges from a process of refinement and increasing formality, with continuous looping back to earlier, less formal stages, as shown in Figure 3.1.

To the software engineer, the design process is closely linked to the refinement of the specification. 'Although the process of setting out a requirements specification as the basis of a contract is a separate activity, formalising that specification may be part of the design process. In

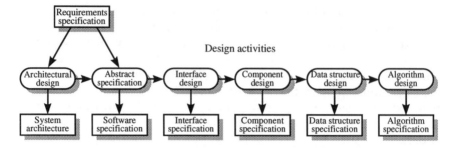

FIG. 3.1 The process of design
Source: Sommerville (1992: 173)

practice, the designer iterates between specification and design' (Sommerville 1992: 173). Thus, the design itself becomes increasingly formalized over time. Because the system requirements and specifications must, at some stage, be frozen so that detailed design can begin, otherwise these become 'creeping requirements' that continuously change, this inevitably means closure whereby the opportunity for inputs to the design process is progressively removed from the users or clients.

Figure 3.2 shows the design activities and the products of those activities. While all these design activities are essential components in the development of large software systems, many software development projects undertake design in an *ad hoc* fashion, often with the design being refined through coding changes (Sommerville 1992). Conversely, structured analysis and design methods, most of which are associated with software gurus such as Ed Yourdon, Michael Jackson, or James Martin, provide formal ways of representing the system using design diagrams, models of system entities and their interrelationships, and data-flow diagrams, etc. Such methods 'are really standard notations and embodiments of good practice' (Sommerville 1992: 176), but their use in British developer organizations tends to be patchy. Even those organizations that identify particular methods and provide training and back-up, find that only a minority of developers chooses to use the method, unless mandated by senior managers or customers as, for example, Structured Systems Analysis and Design Method (SSADM) is mandated for central government projects in the UK.

One way of addressing the design problem, and especially the difficulty of formulating the system requirements and the needs of potential users, is to develop prototype software systems. These provide limited versions of the system enabling gaps and problems to be identified, and alternative solutions to be discovered. Further iterations

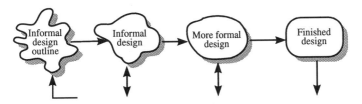

FIG 3.2 Design activities and design products
Source: Sommerville (1992: 174)

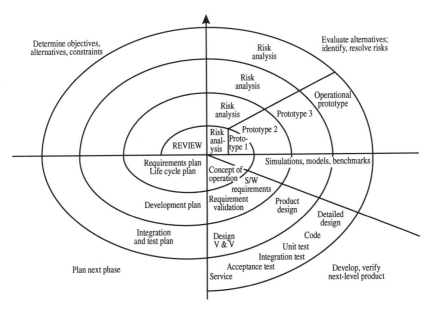

FIG. 3.3 Boehm's Spiral Model of software development
Source: Boehm (1988), © 1988 IEEE

provide more developed prototypes. Boehm's Spiral Model of the development process, including iterative prototype phases, is shown in Figure 3.3.

INHERENT FLEXIBILITY OR ELECTRONIC CONCRETE?

The spiral model of software development has parallels in conventional industrial design and hardware product design, which are also, in some senses, iterative processes. Changes in design are incremental, conventionally leading to generations of products whose lineage can be traced back showing how families of designed artefacts relate to each other.

This is the case for the information and communication technologies that are targeted to consumers, as we saw in Chapter 2, and it also applies to the large technical systems that comprise the communication network infrastructure, which is the focus of the analysis in subsequent chapters. These technological systems are characterized by a degree of path dependency, but as we illustrate below, they also are characterized by varying degrees of freedom in the extent to which their designs are perceived by users to be malleable.

Different types of technological artefacts vary greatly in the periodicity of redesign, and the lifespan or career of each generation. Whereas consumer electrical goods have design life-cycles of a few years, power stations, or wide-bodied aircraft such as the Boeing 747, must recoup investment in their basic design over two or three decades. Even within the same industry there is considerable variation in the redesign cycle. Japanese car manufacturers keep each car design in production an average of four years, while Western car lifetimes average almost ten years (Womack *et al.* 1990).

The concept of generations of designs and their associated careers is perhaps not as clear-cut as this periodic model change suggests. Within automobile model design life-cycles, there will be many minor revisions to specifications, including some minor engineering design changes, most of which will be introduced largely unnoticed. Some may be bundled together as a facelift model which may also have minor styling changes. Conversely, complete model changes that involve a new body design only very rarely do not carry forward existing mechanical components such as engines, transmissions, suspension systems, and many minor components. For the customer, successive new designs are generally made available only to new purchasers—existing customers have to make do with the previous specification, or trade in for the new model.

In the case of software, continuous iterative design and redesign is the norm. However, software products vary greatly in how successive design iterations are achieved and how these are made available to customers. In the case of many types of software, particularly the information systems that support the functions of all types and sizes of organizations, the development process does not stop when the software item is delivered to the customer. Although the development team may hand the software over to the users and the 'maintenance' team, redesign may continue, in different forms, throughout the product life. This continuous process of design is not only the design-in-use effected by users discussed in Chapter 2, it is a continuing involvement of software professionals in the post-delivery maintenance of existing systems.

Software vendors have many options. Volume producers of packaged

PC software launch new releases of their existing products in two- or three-year cycles. In between major releases they may produce interim releases or upgrades. New releases are generally distributed as disks containing the software which replaces the old version. Interim upgrades may be distributed as patches that replace only a part of the existing version. Early versions of new releases of packages are generally sent out to existing and potential customers to expose the software to real world use and gather feedback on problems which may be resolved before product launch.

Where software is sold and supported by a producing firm that has close contact with its clients, upgrades and patches may be transferred regularly from developer to customer. In the *beta*-test phase of a new product, patches and upgrades may be transferred on a daily basis to beta-test clients as bugs are found and eliminated. Because of the ease with which modifications to software can be made, it is essential for developers to keep track of changes and thereby to know at any point in time which version of the product is the definitive version. In this situation, version control becomes crucial. After this initial period, in which thousands of changes may be made in a period of two or three months, the software settles down into maintenance mode. User feedback amasses lists of required changes which are generally accumulated and incorporated into new versions of the software to be released periodically.

Where software is custom-built for a single client, including those many cases where it is developed by in-house developers for corporate users, the acceptance testing phase, also known as *alpha*-testing, will continue until the developers and their client agree that the software meets the requirements as specified. The software then passes into the maintenance phase. Maintenance includes adaptive, corrective, and perfective modifications to the software post-delivery. Table 3.2 defines these and shows the percentage of effort devoted to each as found by Lientz and Swanson (1980).

Estimates of the levels of total software effort devoted to maintenance vary according to the type of system. In the office data-processing environment, maintenance effort is approximately equal to development effort (Lientz and Swanson 1980). In a paper appropriately entitled 'Maintenance is a Function of Design', McKee (1984) reports that between 65 and 75 per cent of total effort is devoted to maintenance. In this sense, software design is an ongoing process that continues even after system delivery. However, the scope for change in the maintenance phase is severely constrained by the early system design decisions. Best-practice system design aims to maximize future flexibility by building in the potential for future modification, and by rigorously documenting system design and components.

TABLE 3.2. Maintenance typology and effort in electronic data processing environments

Maintenance types	Definition*	Effort % of maintenance**
Corrective	System repair: fixing the errors that emerge after delivery of the system	17
Perfective	System improvement: changes to improve the system without changing its functionality	65
Adaptive	System evolution: changes to take account of changes in the system use-environment	18

Source: *Sommerville (1992); **Lientz and Swanson (1980)

Note: **data are from the office electronic data processing environment

The definition of a particular software product or system is thus rather more complex than is the case with hardware technologies. The software in use at any one time is a version of a program that is in continuous development. Where in-house developers are supporting in-house users, system updates might occur very frequently, the replacement software being transferred via a local area network.

The ability to redesign software when it is in use is essential whether the software is custom-built or a standard package. In custom-built information systems, the maintenance of the software will include continuous, lifelong development in order to remedy shortcomings and provide additional functions for users; in the case of the standard package, these will be updated with new releases. In addition, software is increasingly designed to provide the individual user with facilities enabling the customization of the package to suit their requirements.

The malleability and post-delivery adaptability of software is not infinite, however. The scope for change reduces over time; for example, it is relatively easy to change a design specification before coding commences, but not after coding and installation. Initial system design decisions constrain the degrees of freedom open to subsequent developers. Furthermore, highly evolved systems, such as those forming the core of the operations of many large organizations such as banks, insurance companies, and airlines, that have received continuous development over many years, are proving to be a legacy of 'electronic concrete' to many organizations. The sunk investment in such legacy systems, together with their complexity, means that organizations find it difficult to replace them or to make significant changes in their information and communication technology strategies. For example, it may not be possible to gather management information across a number of applications that were developed independently at different times. Additionally, legacy systems mean that a skills base must be retained to support maintenance work using languages and techniques that would otherwise be regarded as outmoded.

There is, therefore, a contradiction between the inherent flexibility and modifiability of software, and the electronic concrete phenomenon. Design and development processes from previous generations of developers are themselves embedded in current-use software, constraining, or enabling, the degrees of freedom available to current developers and to users.

THE DESIGN OF SOFTWARE WORK

This contradiction raises questions about the management and control of the software development process. Just as software products both enable and constrain certain patterns of use, the software development process itself, and thus software work, is the subject of design and redesign. Computer scientists and software engineers seek to impose formal methods of development on software practitioners. Some aspects of developer activity are now being automated, changing the labour market, skills requirements and productivity, and employment opportunities. Whereas the software profession has, for many years, been noted for the autonomy developers exercise in their work (Friedman 1989), recently there has been a backlash in the form of bureaucratic management control. A major effort has been put into making the development process more formal, repeatable, measured, and accountable.

Our research suggests that there are three main trajectories of innovation in software and systems development practice: (i) technical change including new techniques, tools, or methods; (ii) organizational and managerial change associated with quality management regimes; and (iii) commodification or the substitution of generic products or packages for custom development (Quintas 1994*a*, *b*).[7] This work has shown that dialectical processes are present at the level of technical and organizational change within the professional software developer community, particularly in the processes surrounding the innovation and diffusion of new development technologies: the tools, techniques, and methods of software engineering which aim to transform software development from a craft to an engineering discipline (Quintas 1991*a, b,* 1994*c*). Sofware engineering is an oppositional concept which is contingent on the identification of conventional development processes that are regarded as unsatisfactory. The 1980s' government programmes of research and development (R&D) in software engineering supported technical developments in the UK, the USA, Japan, and the European Union. For example, the directors of the Alvey Software Engineering R&D programme in the UK emphasized that their aim was to support the:

critical need to discard the present *ad hoc* 'craft' practices of software production and to ensure that UK management turns increasingly to capital intensive methods of efficient software production and an engineered approach to reliability and quality. (Talbot and Witty 1983: 1.31)

The majority of the effort in these 1980s programmes went into tools and techniques rather than into the organizational aspects of software development. The design phase has come under particular scrutiny. Formal, structured analysis and design methodologies provide rigorous, often theory-based, techniques as well as imposing structure on the project tasks. Computer-Aided Software Engineering (CASE) tools are available to support all aspects of development; upper CASE is focused on analysis and design; lower CASE on programming and testing. Integrated Project Support Environments (IPSEs) and Integrated CASE tools (i-CASE) aim to provide seamless support for the whole software life-cycle. These generally have a range of integrated tools and a central database or repository which support teams of developers with project management, change and version control, and a record of the project history. Their effect is to shift effort away from programming and testing towards user requirements analysis and specification. The most advanced i-CASE tools and IPSEs are based on structured methodologies, and support reuse of specifications as well as code. Within bounded development environments, i-CASE is now delivering significant increases in developer productivity.

Our research also has considered the implications of software engineering tools and methods for the control and autonomy exercised by software personnel, not only over technical choices, but also over the organization of work (Quintas 1991*a*). Software engineering, structured methods, and quality assurance attempt, in different ways, to impose order on the process of system and software design. For example, in providing design techniques and notations, structured methods also contain guidelines that may be more or less prescriptive; indeed, their objective may be to provide consistency and standardization across development teams so that different designers generate similar designs from the same specification (Sommerville 1992). Embedded in CASE tools, structured methods may be made mandatory for developers—enforced by the CASE tool containing a timed record of project history, although CASE tools vary in the extent to which they enforce certain practices and patterns of work. Software engineering based on rigorous methods has implications for developers' work autonomy, skills, and careers. Social and cultural factors may control the pace of adoption of software engineering tools and methods (Quintas 1991*a*). In the crucially important packaged software companies, there is evidence that structured methods are not being applied to the design process. Of 15 package companies in the USA which were studied in detail, Carmel (1993: 85) found that 'most were allocating relatively little effort to formalizing specifications . . . (or) design'.

In recent years, government policy has emphasized quality management standards such as ISO 9001 and BS 5750. Such standards are concerned with work design—its organization and management—rather than development techniques, and they have important implications for the technical systems in which they are embedded, a topic we take up in detail in Chapter 6. Quality assurance issues cover control, documentation, checking, and signing off work packages, but do not address issues of technical practice such as techniques, tools, and methods employed. Quality accreditation schemes provide a framework within which developers apply their tools and methods. Their purpose is to impose transparent process management control intended to leave an audit trail for all development activities including design. In a sense, this is a case of the 'bureaucrats strike back'—whereby management and clients can demand conformance to procedures.

Innovations in technical practice and quality assurance are not the only ways to tackle software development problems. Another important strategy is for user organizations or their contractors to develop new systems by buying in standard software products or packages in place of custom-built software. In this sense, system building-blocks are getting larger through a process of commodification or the use of generic or packaged software (Quintas 1994*a*, *b*) which may improve productivity and reduce uncertainty in development timescales and costs. Use of a tried and tested product reduces risks and ensures a predictable level of reliability and known quality, or, at least, known faults. However, to meet the specific requirements of each application context, packages must be integrated with other systems, and, depending on the complexity of the application, for example, the degree to which it is non-standard or strategically important to the user organization, they may require considerable modification or customization. The introduction of packaged solutions may present difficulties for the adopting organization since they embody built-in assumptions about patterns of use and clients' organizational structure, etc. (Webster and Williams 1993).

Given that the majority of packaged software originates in the USA, and given the emphasis placed on quality management in UK policy, it is ironic that the trend towards quality management does not appear to have spread to the packaged software developers in the USA. Notoriously, software developers greatly resent the chore of documentation and bureaucracy. A large sample of software developers in the USA have been found to have *'ad hoc*, possibly chaotic' development processes (Humphrey 1989). Further research has suggested that package vendors in the USA appear to make a virtue out of fostering a non-bureaucratic creative environment: 'many package companies are dominated or influenced by the hacker culture, and hackers are generally averse to many of quality assurance's formalisms' (Carmel 1993: 86). In the same vein, the President of Sun Microsystems was quoted in *Business Week* (1991) as saying '[the development] process is best practised slightly out of control.'

This accords with current understanding of the conventional, that is, non-software, industrial design process, where bureaucratization is regarded as dysfunctional: 'Formalising design in one way or another so that it fits into current management conventions in order to be controlled and measured may be expedient, but it is not functional. It does not accord with the nature or needs of the design process' (Dumas and Mintzberg 1989: 42).

It is clear that there is a longstanding tension between the creativity and informality of design activity in general, and software design work in particular, and the desire to formalize the software development process. The quotation at the beginning of this chapter suggests that because the design process is not inherently a rational one, wholly formal approaches may be unsuccessful. As Sommerville (1992: 172) puts it: 'Design is a creative process which requires experience and flair on the part of the designer. Design must be practised and learnt by experience and study of existing systems. It cannot be learned from a book. Good design is the key to effective engineering but it is not possible to formalize the design process in any engineering discipline.'

DEVELOPER–USER RELATIONSHIPS

Except where users develop their own software, its development is a complex bridging process linking areas of specialized and diverse expertise: the domain of the information technology professional and the domain of the user. At the extreme, software development must bridge knowledge of the binary operations of computer hardware and knowledge of, for example, the day-to-day management of a hospital, bank, retail chain, or social security office. In the case of embedded information and communication technology systems such as machine, vehicle, or process control systems, software links and controls mechanical, hydraulic, and electrical systems as well as providing user interfaces. In most situations, including those where development is undertaken in-house by user organizations, development requires communication among communities with different backgrounds, knowledge, agenda, and capacities to exercise power. These interactions span the life-cycle of systems, conventionally from the establishment of system requirements and specification, through the design of user interfaces, to system acceptability testing, post-installation modifications, and maintenance.

At the project level, different types of users, for example, customers, managers, and operatives, have different forms and degrees of access to developers, and thus opportunities to input to the design process, at different times over the system life-cycle, that is, from conception to death. As is widely acknowledged, there is unlimited scope for misinterpretations and misunderstandings in these cross-cultural interactions. The communication problem in software development between information technology professionals and system users is a well-charted and longstanding issue (Mumford 1972; Murray and Willmott 1993; Robey and Farrow 1982). The problems associated with meeting users' requirements have taken on added importance in the Information Systems area as, increasingly, the user group includes middle and senior management (Friedman 1989). In contrast to the data entry clerk users of the 1960s, these managers exercise organizational power and the freedom to question system design characteristics.

Software which is developed by users for their own purposes avoids the familiar separation of conception, design, and use. Conversely, when systems are being developed by one group for use by another, there is a need to consider the relationships between developers and users, and issues such as prefiguring and closure. There are very large variations in these interrelationships. Designing a one-off information system tailored to the needs of a single large organization has quite different opportunities for involving the users as compared to the design of a software package that will be sold across the world in volumes measured in millions.

The software community has devoted considerable effort to tackling the perceived problem of the communication gap between the designers/developers and the system 'users'. 'Users' is placed in quotation marks here in acknowledgement of the fact that it is a problematic concept.[8] In the conventional information systems software development life-cycle, designer–user interactions occur first in the requirements analysis phase. The software developers then design and build the system as specified as a result of requirements analysis. Requirements analysis is an interpretative process; the people who purchase and use information and communication technology systems must find ways of communicating their needs to the professional software developers; the latter must attempt to ascertain what the 'users' might want, in the light of what they, the technical specialists, know to be possible. However, the 'users' that interact with the systems analysts and designers may not include the 'end-users' who will directly use the finished system. The design interactors are often representatives of management and system procurers rather than end-users, although they will seek to represent user 'needs'.

Even when software is developed in-house by user organizations, the distinction between developers and users still invariably applies. The fact that software design and development is undertaken by in-house personnel does not inevitably mean that the developers are better linked with, or more sympathetic to, system users or potential users than an external contractor would be. The history of computing shows that in-house information technology departments have not always achieved close working relationships with their internal client groups (Friedman 1989). Although some organizations have experimented with close user involvement in software design (Hales 1993), and even user-led systems design (Franz and Robey 1984), such innovations undoubtedly remain exceptional.

In the systems and software engineering literature the main user issues addressed are concerned with two areas. First, there are the difficulties associated with establishing the user requirements and translating these into a system specification. The disproportionately high cost of getting this wrong is emphasized (see, for example, Boehm 1981). The second user issue concerns the development of an appropriate user interface or front-end for the system (Sommerville 1992). There are a number of technical solutions applied to the developer–user communication problem and the inevitable 'creeping' requirements during requirements analysis, design, and development. One strategy is to build adaptability into the software to ensure that it can be made responsive to users' changing requirements after implementation. Another strategy focuses on the development of a series of prototypes that can be demonstrated to users and provide the basis for iterative designs which evolve towards a full system (see Figure 3.3 above).

However, these technical processes cannot be divorced from the social contexts in which they take place. Hirscheim and Klein (1989) have pointed out that information system developers approach requirements analysis and design issues from particular perspectives and make choices on the basis of particular assumptions. While the dominant developer paradigm is essentially functionalist, with the systems analyst assuming the role of technical expert employing instrumental reasoning in pursuit of the 'true' system requirements, this rational approach often fails to produce appropriate systems. Further, technical rationality is only a partial reflection of the range of underlying developer assumptions that influence the course of developer–user interactions.[9] The system that results is not so much an objective representation of 'true' user needs, as the result of particular social relations and the exercise of professional power (Markus and Bjørn-Andersen 1987).

Power does not just reside with the software professionals. Those paying the bill may impose their own views on the design process, or impose financial constraints, which conflict with the needs of the real system users. This is partly what happened in the case of the development of the London Ambulance Dispatch System that brought the capital's emergency service to a halt in November 1992. The people who ran the existing manual system were not consulted by the developers about their procedures and tasks—the ambulance system's management made assumptions about those tasks and practices and relayed these to the developers (SWTRHA 1993).

What of developer–user interactions in the implementation and maintenance phases? After system build and testing, the software is installed in the user organization. It is at this stage that developers may well find that the initial specification was inadequate, and that the requirements of the users have changed over the period since the requirements analysis phase. Moreover, the requirements of the users are likely to be changed by the introduction of the new system. This is not just a software problem. Leonard-Barton (1988) has suggested that the implementation problem observed in the introduction of new process technologies across a range of conventional industries and technologies, may be seen as requiring successful management of the process of mutual adaptation of technology and (user) organization, that is 'the re-invention of the technology *and* the simultaneous adaptation of the organization' (Leonard-Barton 1988: 253). Here, a mismatch between the characteristics of the process technology and the performance needs of the user organization is assumed to be the norm. Leonard-Barton (1988: 265) concludes that 'implementation *is* innovation'.

Many software systems are long-lived, and the people involved in maintenance on each side of the user–developer divide are likely to be different groups from those involved in systems development. The users will include end-users. The software professionals are likely to be a designated maintenance team. Both groups have to require or make changes within the constraints of the existing software system. However, maintenance activity is not simply shaped by technical factors embedded in the existing software. Kling and Iacono (1984) suggest that the control of post-implementation development is a function of organizational politics. Systems are 'pushed in a specific direction which would increase the power and control of key actors within the organization' (Kling and Iacono 1984: 1225). In this case, the key actors were not the software professionals, but rather were senior manufacturing managers.

The history of user–developer interactions is one characterized by many changes of fortune. The need to consider the communication problems that inhibit interactions between software specialists and user representatives, and power relationships regarding the control of specialized knowledge domains, were recognized in the early days of information and communication technology application in user organizations (see Kozmetsky and Kircher 1956). The history of structural power asymmetries in information and communication technology development within user organizations shows an initial concentration of power within centralized Electronic Data Processing (EDP) departments. Power resulted, at a structural level, from the concentration of expert knowledge and the location of EDP units within finance departments. However, there are many other ways in which software professionals have exercised power over users (Markus and Bjørn-Andersen 1987). At the level of social relations, software professionals' use of specialist language and appeals to a 'technical' realm were, and indeed still are, used to circumscribe areas of knowledge and technical expertise that delimits and controls user–developer discourse (Bloomfield and Vurdubakis 1994; Low and Woolgar 1993).

The structural basis of power was partially broken in many organizations by the availability of microcomputers in the late 1970s, and PCs in the early 1980s. Many micros found their way into offices, schools, and other workplaces, with home-computing enthusiasts writing some of the application software. Purchases often were made out of petty cash or other budgets to avoid the need to involve central computing departments. The subsequent availability of word-processing and spreadsheet packaged software made the machines more useful and accessible to a wide number of users. For a short period in many organizations, the central EDP departments lost control of a rapid and unco-ordinated proliferation of PCs and software. Ironically, control of the design process rapidly slipped away from the amateur departmental programmers and the EDP departments to the suppliers of packaged software who were invariably far more remote and inaccessible to users than the EDP department.

The information and communication technology professionals have largely regained control of organizational information systems development with the spread of networks, centrally stored PC software, for example in client-server systems, management information systems, and the need to develop systems that interface with other organizations;

all activities that have to be co-ordinated centrally. Furthermore, developer effort has shifted towards higher-level business analysis. The concept of business process re-engineering, whereby organizational processes are restructured in order to align working practices with the best practice characteristics of information and communication technologies, aims to place information and communication technology developers at the centre of organizational development. Rather than designing the information and communication technology system to fit the needs of the organization, or, as Leonard-Barton suggests, mutual adaptation, this approach proposes that the business processes of the organization be changed to best fit or exploit the characteristics of information and communication technology systems (Scott-Morton 1991).

END-USER PROGRAMMING

In the early days of telephony there were suggestions that the spread of the technology would reach limits imposed by the number of operators required; these 'would soon reach ridiculous levels because everyone in the country would have to become an operator. Well that's exactly what happened. In a sense, each telephone user is an operator, able to achieve some degree of end-user programming' (Booch 1994: 37). As we show in Chapters 4 and 5, however, there are very real questions as to the utility of many forms of user-initiated programming. These concern the degrees to which communication network users, for example, are able to design networks that achieve a degree of closure or exclusion for others, and the extent to which a measure of privacy can be maintained in public electronic communication environments.

We have noted that the arrival of microcomputers and PCs in the late 1970s and early 1980s provided end-users with the possibility of developing their own usually very limited software applications. Such end-user computing was driven by the problem of backlogs in central provision of applications, exacerbated by the problems surrounding the interactions between professional developers and users. This form of end-user computing was essentially a covert and spontaneous response by users. There also have been many overt attempts by information and communication technology suppliers to provide end-users with products intended to enable them to program their own systems, thus bypassing communication problems, and, in addition, removing the productivity problem from the professional developer sphere altogether. High-level development languages using commands that approximate to natural language originally were billed as providing end-users with the tools to produce their own applications. These 4GLs (Fourth

Generation Languages) generally proved to be too complex for the average user to master quickly, and so the skills associated with 4GLs, such as SQL (Structured Query Language),[10] became part of the professional software developer's portfolio.

Arguably, the most successful examples of end-user programming are spreadsheet products, and the packages based around them, such as Lotus 1-2-3. These do enable persevering end-users to develop small applications on an *ad hoc* basis to support their own needs. Strictly speaking, spreadsheets are user-configurable systems rather than user-development tools, since the options open to the user are heavily constrained by initial design decisions embedded in the package. End-user programming has thus shifted away from the idea of the user as programmer, putting great effort into mastering fully functional programming languages and writing lines of code. Instead, end-user programming is based on a process of commodification, whereby the user purchases a software product such as a spreadsheet package that supports a self-service application development path, analogous to the way in which domestic services have been replaced by purchased commodities (Gershuny and Miles 1983).

R&D on user-configurable systems as a generic approach to software development has received support from the UK government. So, too, the idea of building systems from reusable software components has been pursued for over a decade. At present, reuse of components by software professionals is not widely practised, in part, because of its potential impact on software work practice. Reuse is likely to lead to large productivity gains if the problems associated with accessing libraries of components and integration can be overcome. Whether the trend towards reuse of software modules and objects will enable end-users to design and build systems out of purchased software components that may be assembled into complete systems, is open to conjecture. It is possible that end-users may be more receptive to the principle of reuse than are software professionals. However, overall systems design will remain crucial, and is likely to continue to require highly skilled professional expertise. There is no obvious solution to the fundamental problem that users, by definition, only rarely develop systems and thus lack the cumulative experience that professional developers who develop system after system are able to amass.

SOFTWARE LIFE-CYCLES AND INNOVATION

This chapter has emphasized the need to examine software design decisions throughout the software life-cycle, from system conception to retirement. We have seen that different design interactions influence developments at different stages. The need to acknowledge hetero-geneity has also been shown to be paramount. Andrew Friedman has suggested that the locus of information systems development has moved inexorably further away from machine-level software towards the end-user and the application domain (Friedman 1989, 1993). High-level languages, end-user computing, and systems development methods that seek to model business processes, are all factors that support this analysis. However, this progression conceals a counter-trend that has come to the fore in our research and which has significant implications for the locus of design. What we have found is the large-scale substitution of standard packages for custom design and development, both at the PC and information systems levels. Such packages remove key design decisions from local developers. It is crucial, therefore, to differentiate between initial design decisions as embedded in standard products, and subsequent design decisions in so far as these are enabled or constrained by the initial choices.

It is widely accepted that technology is socially constructed; that is, that innovation takes place within economic, social, and institutional frameworks; technology embodies certain values and assumptions which have implications for users, predetermining use-patterns. However, as pointed out by Orlikowski (1992), accounts of technological development and use that assume unidirectional causal processes are flawed. Orlikowski suggests instead that these processes are better conceived as those in which technology is socially constructed but also reinterpreted in use as we have discussed in Chapter 2. Use patterns, in turn, depend on the circumstances in the user organization (Wilkinson 1983). In practice, technology is shaped by the social setting in which it is both developed and used (Edge 1988; MacKenzie and Wajcman 1985). So, too, users subvert technology, using it in ways not envisaged by the designers, effectively designing-in-use.

Technologies vary in terms of how far they are flexible in use—how far they prescribe and limit the patterns of use, and the organizational forms that support their use, within a work or leisure setting. The simplest technologies, for example, a hammer, prescribe little, while many forms of automated and electromechanical machinery tend to be highly prescriptive, for example, power looms and automated transfer lines in mass production systems. This is particularly the case with systemic technologies, that is, those requiring managed teamwork.

There is a spectrum of potential to redesign technology-in-use. Pinch and Bijker (1984) refer to the interpretive flexibility of technology or the extent to which users are able to influence its attributes either in design or use. It is important to acknowledge that technologies vary in the extent to which they prescribe use-patterns and constrain future degrees of freedom. So, too, there is a time dimension; user–developer interactions change over the career or life-cycle of a product. In the case of software, it is essential to distinguish between design decisions that are embedded in standard packages, and the subsequent degrees of freedom available to downstream integrators and users of these packages.

A central premiss of this chapter is that information and communication technology is different from other conventional technologies because of its inherent informational characteristics, and particularly because of the flexibility of software and the structure and form of software design and development capabilities. The versatility of computer-based systems derives from the separation of hardware and software which enables standard hardware to be reprogrammed. Software tailors hardware for a multiplicity of uses. In addition, much software, particularly information systems in organizations, is modifiable even after delivery, maintenance activity being a process of continuous development.

The key question is how far do predelivery design decisions constrain or enable use patterns and future developments? The issue is not whether system design captures the 'real' requirements of the real system users during development, but how far it builds in the flexibility for future development in the directions likely to be needed. We have noted the contradiction between the inherent flexibility and modifiability of software, and the electronic concrete phenomenon of sunk investment in legacy systems. Such systems constrain the degrees of freedom available to users and to post-delivery developers. Developers seek to design systems that will support future evolution and development, but it remains to be seen whether or not fundamental design decisions will always provide a limitation.

CONCLUSION

The structure of software development capability is unique, principally because of the lack of concentration of that capability within a supply-side, hypothetical software industry. The locus of software design and development has changed continually over the last fifty years. During these transitions there have been shifts in the balance of power between suppliers, developers, and users. However, having emphasized that software development is different from other technological activities, it is also clear that, in some respects, for example, in the widespread diffusion of standard products leading to the dominance of the software supply-side, software is becoming more like conventional industrial activity. The capability to develop software remains highly decentralized, but the software professionals located throughout the economy are a diffuse resource that is under threat. In particular, standard bought-in packages are substituting for locally custom-built systems. The design and development capability at the user-organization level, representing a highly devolved and unique ability to innovate, is losing ground to the burgeoning growth of systems based on product diffusion.

Any reduction in the plurality of innovative capacity will have consequences for the degrees of freedom available to software users. The packaged software market is now in the hands of a few powerful organizations whose impact is greatly disproportionate to their size. The view that PC software diffusing around the globe can be seen as a form of cultural imperialism, imposing new modes of working on millions of users who have no input into the design process (Quintas 1994*d*), might be regarded as overreaction. However, the idea that software *is* culture, or, more specifically, that packaged software embodies US culture, has been proposed as an explanatory factor for the competitive success of US software products world-wide (Carmel 1994: 11–12): 'The US packaged software industry is composed of a number of behavioural factors which, together, create a uniquely American flavour to the industry and its products. This cultural flavour restrains foreign competition.'

Our understanding of software has implications for wider industrial design, innovation, and manufacturing processes. Through their centrality to high value added economic activity, design and innovation assume ever greater importance. Software design and development may be expected to provide insights into emergent innovation processes. On one level, the application of software and information and communication technologies is transforming the design to manufacture process. For example, software-intensive telecommunication systems are enabling the transformation of customer–supplier networks, including electronic data interchange, value added network services, groupware, etc., and some of these developments are considered further in Chapters 4 and 5.

On another level, the software model of technological design can be seen as an archetype for a more general emerging pattern of industrial activity (Rothwell 1992). In this sense, conventional hardware-intensive industries are becoming more like software. Conversely, and ironically, the trends towards formalization and bureaucratization of software development, plus the ever-increasing importance of the software package suppliers and substitution of packages for custom building, suggest that software design, development, supply, and use are now becoming more like traditional industrial processes.

NOTES

1. These data are for information technology and do not include telecommunication equipment and services.
2. Examples include fuzzy logic software in automatic transmission systems on BMW and Mitsubishi cars.
3. We only briefly look at the subject of innovation within the software development process, (see Quintas 1994*a*, *b*).
4. The gender-specificity of the imaginative designer belies the fact that the authorship of the first program conceived to run on Babbage's Analytical Engine is attributed to Lady Ada Lovelace, *née* Augusta Ada Byron (Lord Byron's daughter) who is now recognized as the first computer programmer (James and Morrill 1983).
5. As well as the instruction manuals and other documentation that go with these.
6. Pavitt's (1984) analysis differentiates between sectors according to the determinants and nature of their patterns of technical change. Supplier-dominated firms have low internal technological capability. Science-based firms are R&D intensive and produce a high proportion of process and product innovations. Production-intensive firms exploit economies of scale and maintain high levels of innovative capability. Specialized suppliers provide equipment to specific markets, maintaining innovative capability for both process and product innovation geared to customers' needs. Although empirically derived, the analysis does not deal adequately with software development and innovation.
7. In this chapter, we use the concept of commodification to mean the substitution of a packaged bought-in software product for developer activity. The software package displaces developers' labour, effectively substituting a standard commodity for design and development activity.
8. Friedman (1989) has identified six categories of user: (i) the patron, initiator, and champion of the system; (ii) the client, for whom the output is intended and who will pay the bills; (iii) design interactors, that is, people involved in system specification and design; (iv) end-users, the individuals who will manipulate the installed system; (v) those involved in maintenance and enhancement of the system; and (vi) secondary users, including people displaced by the new system, people whose work is changed by the system, those whose non-work life is affected, and proxy-users such as departmental managers and union representatives.
9. Hirscheim and Klein (1989) suggest that there are four paradigms of information systems development—functionalist, social relativist, radical structuralism, and neohumanism. Associated with each of these is a developer archetype: technical expert, facilitator, labour partisan, and social therapist.
10. SQL is the Structured Query Language, one of the most successful 4GLs, used in conjunction with relational databases.

4

Designing Electronic Commerce

ROBIN MANSELL

The ability to overcome space is predicated on the production of
space . . . [there is tension] between fixity and motion, between the
rising power to overcome space and the immobile structures
required for such purposes.

(Harvey 1985: 149)

INTRODUCTION

Electronic commerce refers to the ways that firms are engaging in the
conduct of business by using advanced electronic information and
communication networks. The conduct of business in this way requires
the conjuncture of technical and institutional design features as well as
new capabilities on the part of firms to enable them to construct,
manage, and use rapidly expanding global electronic networks.
Electronic commerce enables business to be conducted at a distance.

Harvey (1985) has recognized that the process of overcoming the
constraints of geographical space, which involves the dematerialization
of many kinds of exchange relationships, also necessitates the involvement of relatively fixed structures. It is the relative fixity and flexibility of
both the technical and institutional aspects of electronic commerce that
determine whether firms are able to use electronic commerce to
competitive advantage. The dialectic that characterizes the innovation
process which brings increasing reliance on electronic networking in the
conduct of business activities involves four main domains of activities.
Each is crucial to the design and maintenance of networks.

First, electronic commerce is feasible only to the extent that an
appropriate network infrastructure or substructure is in place. The
design of this substructure (including the switching and transmission
components comprised of hardware and software) is increasingly
characterized by 'intelligent' software that is configured with computer
and telecommunication hardware to create the complex networks that
span the globe. Second, electronic commerce depends on the design of
services which may include any combination of voice, text, and image-
processing. Third, electronic commerce is initiated by firms whose

actions are influenced by the environment in which they conduct their business affairs. This environment includes policies and regulations that govern the design and use of the technical substructure and services as well as the actions of firms that are both competitors and collaborators. Finally, electronic commerce is shaped by the organizational character-istics of firms and their capacity to learn to creatively produce and use the network substructure and advanced information and communica-tion services.

In this chapter, we outline some of the tensions that affect the design of the technical substructure and the services that are the components of electronic commerce. We look particularly closely at one facet of electronic commerce—electronic trading networks—in order to illustrate the interdependencies between the four critical domains of electronic commerce. The aim of the chapter is to consider the challenges that are presented to firms as they interact with designers of networks to meet their changing business requirements. We consider the extent to which electronic commerce is enabling firms to modify the constraints of time and space as they gain access to public telecommun-ication networks that are more flexible and 'intelligent' than they have been in the past. The principles of design and capabilities are employed to examine the resources that firms bring to the management and operation of electronic commerce networks.

FIRMS, INNOVATION, AND ELECTRONIC COMMERCE

Electronic commerce is associated with the dematerialization of trans-actions into electronic space, but it involves much more than this. Complex technical and organizational systems must be in place in order for electronic commerce to bring advantages to firms. Intricate sets of relationships need to be established and these evolve unpredictably. Software has characteristics that are both malleable and analogous to 'electronic concrete' as we saw in the preceding chapter. Software provides the 'intelligent' functionality which supports today's networks of electronic commerce. As is the case for software applications in other areas, the design of communication networks is strongly influenced by the legacies of past designs and investment patterns. These legacies affect the way in which the network substructure and services can be mixed and matched to support the information and communication requirements of firms. They also affect the degrees of freedom available to firms to alter the way they engage in information management and their knowledge production and consumption activities. As Dosi has argued, as firms learn to develop and use complex technological systems, they build on their existing knowledge base and other assets. Thus,

[T]he search process of industrial firms to improve their technology is not likely to be one where they survey the whole stock of knowledge before making their technical choices. Given its highly differentiated nature, firms will seek to improve and to diversify their technology by searching in zones that enable them to use and to build upon their existing technological base. In other words, technological and organizational changes in each firm are cumulative processes too. What the firm can hope to do technologically in the future is heavily constrained by what it has been capable of doing in the past. (Dosi 1988: 225)

However, choices of firms with respect to electronic commerce networks are not wholly constrained by the past. In varying circumstances, the design of these networks becomes malleable. A change in any one or more of the four domains of electronic commerce can lead to both intended and unintended outcomes which affect the ability of firms to compete successfully in local or global markets.

The literature on the business use of information and communication technologies is burgeoning and increasing attention is being given to the innovation process in the organizational and technology management domain. The problems encountered in developing and implementing inter- and intra-firm networks are the subject of numerous case-studies, as are the factors contributing to success and failure.[1] The insights from research range from Porter and Millar's (1985) prescriptions as to 'how information gives you competitive advantage', to the social constructivists' accounts of the processes whereby shared understandings are negotiated among network and information systems suppliers and users about how the technical artefacts which comprise these systems should be designed, or 'metabolised' to use James Fleck's (1994) expression. Yet this growing body of research has yet to address many of the transverse linkages between the technical and institutional domains that influence whether electronic commerce will succeed in bringing benefits to the majority of potential user firms.[2]

The Office of Technology Assessment (1994) in the USA uses an 'electronic commerce matrix' to relate alternative modes of communication (point-to-point, point-to-multipoint, and multipoint-to-multipoint) to transaction costs (information search costs, exchange costs, and monitoring costs). As Figure 4.1 shows, the cells of the matrix indicate how the communication infrastructure provides the medium for the numerous services that support the information and communication requirements of firms. For example, on-line databases, audiotex, and videotex services are configured to be responsive to the need to reduce information search costs and, generally, they are provided via point-to-point modes of communication. Point-to-multipoint applications such as telemarketing and teleshopping are used to reduce the costs of completing transactions in a market. All the service applications shown in the matrix depend on the existence of a network substructure which

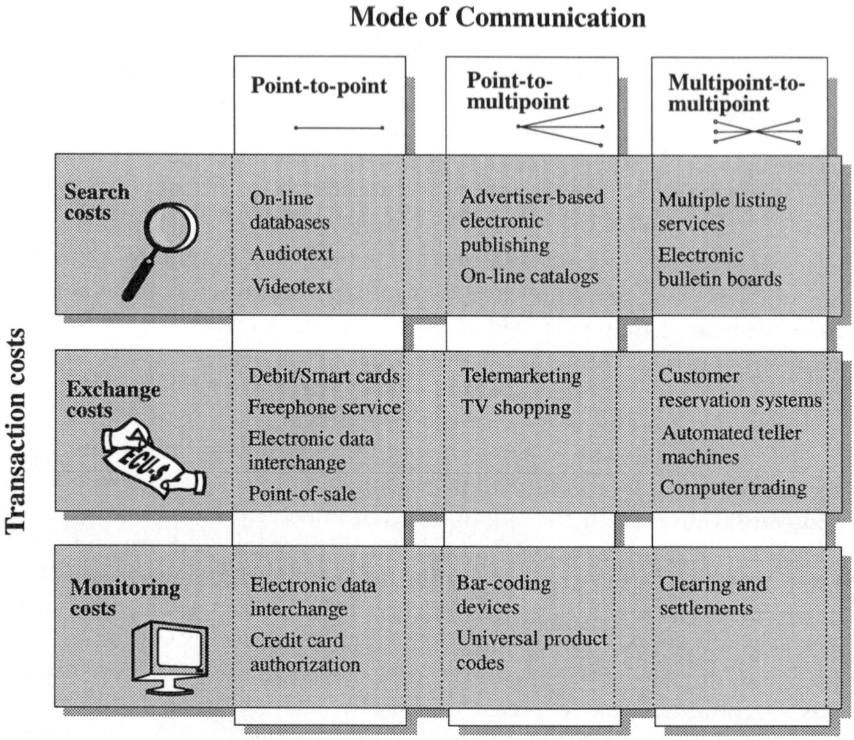

Mode of Communication

	Point-to-point	Point-to-multipoint	Multipoint-to-multipoint
Search costs	On-line databases Audiotext Videotext	Advertiser-based electronic publishing On-line catalogs	Multiple listing services Electronic bulletin boards
Exchange costs	Debit/Smart cards Freephone service Electronic data interchange Point-of-sale	Telemarketing TV shopping	Customer reservation systems Automated teller machines Computer trading
Monitoring costs	Electronic data interchange Credit card authorization	Bar-coding devices Universal product codes	Clearing and settlements

(Left axis label: Transaction costs)

FIG. 4.1 Electronic Commerce Matrix I
Source: Adapted from Office of Technology Assessment (1994)

can be configured in different ways to include terrestrial and radio-based, fixed and mobile public and private telecommunication networks, cable television networks, and varying degrees of interactivity between customers and suppliers, and network operators and users.

Our approach, in contrast, seeks to understand the changes in firms and markets that are intermediated by electronic commerce, and this requires that we investigate the changes in the relationships among the actors involved in network design throughout the development and implementation processes. In the literature on the increasingly wide-spread diffusion of electronic commerce, there is an assumption that, once in place, this mode of transacting inevitably will bring benefits to firms in the form of enhanced competitiveness. Once we recognize, however, that the design of electronic commerce and the requisite technical and organizational capabilities are not automatically given, it is necessary to look in greater detail at additional aspects of the technical and organizational innovation process. To achieve this it is helpful to examine decisions that are taken at the interfaces between the domains in which actors negotiate and select network configurations and services.

One important issue for the strategic manœuvres of firms and for policies intended to support the wider diffusion of electronic commerce is the extent to which, on balance, firms are learning to design networks and services in ways that take advantage of the constraints and opportunities which appear over time in all four of the domains of electronic commerce. In contrast to the matrix employed by the Office of Technology Assessment in the USA, our matrix focuses upon the four electronic commerce domains outlined above, that is, the network infrastructure or substructure, services, policy and regulatory issues, and the organizational characteristics of firms.[3] Figure 4.2 shows the network domains where conflicts and divergent design considerations arise on the vertical axis. The diagonal axis shows the key information elements relevant to the conduct of business in an electronic marketplace.

There are five main information elements in the process of exchanging goods or services in a market whether that market exists in a physical place, is partially dematerialized, or is only present in an electronic environment. Each of these elements involves the production, exchange, and use of information, and a communication process.

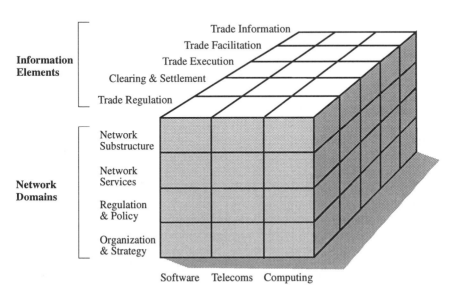

FIG. 4.2 Electronic Commerce Matrix II
Source: Adapted from Office of Technology Assessment (1994)

● Trade information includes the dissemination of background information, the possibilities and options which comprise the knowledge base that is required to become involved in the exchange of goods and services. Examples include databases on patents, standards, catalogues, and information such as market intelligence and news agency reports.

● Trade facilitation includes the information and procedures concerning activities, actions, and choices needed to allow the movement of goods and services through physical or electronic space. This includes legal procedures (customs clearances, health and safety requirements), financial procedures (insurance, letters of credit), logistical supports (tracking and tracing services, certifications, authentication, storage, repackaging), and administrative information (instructions, product or service descriptions).

● Trade execution involves the process of informing relevant parties that a binding contract for the exchange of goods or services has been consummated. This information may take the form of a private agreement in a closed market or an open sale, and the process of execution may be through tendering, auctioning, or a fixed price.

● Clearing and settlement involves all the information-related actions required to move goods or services and funds once an exchange of a good or service has been executed. For example, information about title transfers, delivery, and financial guarantees and funds transfers, must be completed.

● Trade regulation involves information concerning all the elements of exchange required by voluntary or statutory organizations. This may cover rules with respect to transborder data flows, capital adequacy, settlement periods, access to markets, quotas, tariff and non-tariff barriers, intellectual property rights, and other matters.

Each of these information elements imposes different kinds of demands on the design features of the physical substructure and of electronic services. These requirements, in turn, are informed by the strategic positions of firms in their respective markets. Thus, for example, there can be considerable heterogeneity in the levels of network access, quality of data transmission and security, speed of communication, network and service standards and gateway protocols, regulations for domestic and cross-border trading and dispute settlement requirements, that are desirable for firms engaged in different commercial and industrial activities in locations around the world. Different network and service specifications (standards, protocols, pricing arrangements, etc.) are appropriate to the various aspects of information production and use throughout the process of buying and selling in electronic markets. In order for electronic commerce to support the competitive strategies of firms, choices as to technology and rules and codes of conduct must be taken with regard to the technical substructure and services as well as the organizational characteristics (internal and external boundaries) of firms.

The problem of innovating in the field of electronic commerce can appear to be deceptively simple. For example, in the rhetoric that often accompanies the promotion of advanced networks and services, such as electronic data interchange or electronic funds transfer, the capabilities of suppliers and users of the technical substructure and services appear to need only to be matched together to create, process, and distribute electronic information in such a way that the constraints of time and physical space are overcome, thus enhancing the competitiveness of firms. However, the appropriate technical components of networks and the necessary actions on the part of suppliers and users to bring about a successful match between the technical and organizational design domains are difficult to determine and this is especially the case as networks extend their reach and traverse the boundaries of countries and regions. The effective design of electronic commerce networks is characterized by considerable uncertainty; uncertainty that is generated by changes in the technical aspects of the substructure, the functional specification of services, the competitive environment—locally, regionally, and globally—and disparate, and often inconsistent, policies and regulations.

Over the last decade firms of all sizes have begun to implement systems of electronic commerce. This requires that they access and use the telecommunication infrastructure and the growing number of service applications this infrastructure supports. The information highway vision that is being promoted extensively in the 1990s offers an impression that firms will simply be able to 'plug' into the substructure and its services and proceed to design electronic environments that suit their business needs. Electronic commerce, in its various forms, is expected to offer enormous time- and cost-savings in the conduct of business in both the manufacturing and services sectors. However, as von Tunzelmann (1994: 20) argues, the shift from mechanization to 'informatization' based on information and communication technologies has coincided, historically, with an 'escalation of complexity of use'. It is therefore in the intersection between the complexity of network use as experienced by innovating firms and the complexity of network supply that we should expect to find insights into the implications of electronic commerce.

Electronic commerce involves communication and the exchange of information and, as Coase (1993*b*: 233) has observed, 'it makes little sense for economists to discuss the process of exchange *without specifying the institutional setting* within which trading takes place since this affects the incentives to produce and the costs of transacting' (emphasis added). The institutional setting for electronic commerce is particularly significant in determining why firms behave as they do. The neoclassical economics approach to this issue treats the institutional setting which leads to the selection of technologies and the organizational character-istics of firms as a 'black box'; little consideration is given to how and why firms acquire the capacity to initiate the actions which shape their technical and organizational environments.

Research on the advantages and disadvantages of electronic commerce tends to be similarly blind to the specific characteristics of firms and the technologies that enable them to transact some, or all, of their business electronically. Thus, for example, some writers suggest that advanced information and communication technologies can be configured so as to create near-perfect markets, markets which come very close to the characteristics of the neoclassical ideal. In such markets, transaction costs are reduced to nearly zero and a firm's participation in an electronic market is expected 'automatically' to contribute to its success in the market-place (Butler Cox Foundation 1990; Hootman 1972; Kimberley 1991; Malone *et al.* 1987).[4]

Malone suggests, for instance, that electronic commerce contributes to the reduction of information asymmetries on the part of participants in a market as a result of three main effects. Advanced information and communication technology systems increase the amount of information that can be exchanged at reduced cost (communication effect); they enable a better match between buyers and sellers to be achieved (brokerage effect); and they support the more effective coupling of stages in the value added chain (integration effect) (Malone *et al.* 1987). Firms themselves are regarded as 'black boxes' as are the technical configurations they use. Studies of organizational change and strategic management theory are brought to bear on issues concerning the more effective use of electronic commerce, but problems that arise as a result of mismatches between technical designs and firms' internal character-istics and positions in the market, as well as the wider policy and regulatory environment, in most cases, are left to one side.

Our perspective on the innovation process in information and communication technologies requires that we recognize that the technical substructure and services of electronic commerce are no more neutral with respect to markets and the prospects of firms than are any other features of firm behaviour. As Ciborra comments in his analysis of the characteristics of electronic commerce, 'creating an electronic market is equivalent to creating a dance hall, or a veritable market-place or market square, where people can meet, and make a profit out of it' (Ciborra 1994: 108). Just as in any set of proximate relationships where people do not meet on equal terms, firms do not meet on equal terms when they engage in the design and use of electronic environments. Relative disparities in the power to design networks and in the capabilities of agents are the essence of the dialectic of the control and use of networks. As Giddens argues, 'power is the capacity to achieve outcomes' (Giddens 1984: 257), and this aspect is neglected in most studies of the implications of eletronic commerce.

We turn next to an examination of the design characteristics of the public telecommunication network and advanced information and communication services as these are the constraints within which the majority of firms must select the components of their electronic commerce systems. We look first at developments in the substructure and at the two contending visions of network development that are prevalent in the 1990s—the Internet vision and the entertainment vision. Second, we consider several regulatory and policy issues that firms must contend with when they engage in electronic commerce. Finally, we turn to an examination of the resources and capabilities that clusters of firms in different countries and sectors have brought to bear on the creation of electronic trading networks, a distinctive type of electronic commerce, and we consider the degree to which these firms have established relatively exclusive and closed networks in a bid to insulate themselves from the uncertainties of the market-place and to strengthen their competitiveness.

THE SUBSTRUCTURE AND SERVICES OF ELECTRONIC COMMERCE

By the mid-1980s, the public telecommunication network in most countries was beginning to incorporate digital technologies and computerized intelligent switching equipment. The Plain Old Telephone Service which relied on a limited range of technologies was being transformed by a maze of new technologies (see Mansell 1993*a*). A vision of a high-capacity (broadband) network infrastructure or substructure providing access to information and communication service applications running at high speeds and reaching into homes and businesses is being promoted by governments and the private sectors in the major trading regions of the world. In the industrialized countries, many of the newly industrializing countries, and, indeed, in some of the poorest countries, a broadband infrastructure is expected to be in place in the next decade. In some countries, this advanced high-capacity network substructure is expected to replace the existing infrastructure which is comprised mainly of relatively low-capacity copper wire or coaxial cable (narrowband) when it reaches into most homes and businesses. In other cases, the price—performance characteristics of optical fibre and digital radio communication hold out the promise that those who have been excluded from today's advanced networks of electronic commerce will be able to gain access to broadband, or to upgraded narrowband, networks within the next decade. Major construction programmes are being planned in most countries and countries are vying with each other to proclaim leadership in the race to introduce the next generation hardware and software of electronic communication.

In technical terms, there is little to differentiate the visions current in the USA, Europe, Japan, and many other countries. In the USA, the vision is articulated along the following lines:

[S]imply put, fibre to the home, school and business is an essential infrastructure for economic development in the Information Age of the 21st century, just as railroads were in the last century and highways were in this century . . . The installation of fibre and the development of broadband networks . . . will fundamentally improve our international competitiveness in the Information Age because it will bring new and different services with it that will improve our national productivity and our quality of life. (United States Senate 1990: 27)

In Europe, similar ambitions prevail: 'The digital revolution will prompt a reorganization of key industrial sectors and the emergence of new services which have the potential to transfigure the European society and the daily life of European citizens' (Group of Prominent Persons 1994: 1).

Although the architectural design of the new networks is considerably advanced, the rate and timing of investment in these information highways is likely to be informed by the particular conjuncture of the elements in our 'matrix of electronic commerce'. The designers and users of the network substructure are not 'lightening calculators of pleasure or pain' as Veblen long ago observed, and the trajectories of innovations in the technical elements of these networks become especially uncertain when they move into the realm of commercialization and use on a widespread basis. In addition, the upgrading and implementation of the technical components of networks are affected not simply by the apparent technical superiority of one generation of hardware or software over another, but by the resolution of conflicts among interested parties in the supplier and user communities, parties with unequal resources and capabilities to advocate new network design configurations.

Nevertheless, and in spite of uncertainties in the supplier and user communities, it is generally expected that future broadband networks will be established using fibre-based local area networks, fibre data distributed interface systems, and metropolitan or wide area networks as transmission facilities. Digital voice, data, and image traffic is expected to grow when bridges are established between physically distinct networks. Bandwidth 'hungry' applications such as computer-aided design and manufacturing and video services, together with the build-up of traffic generated by applications such as electronic data interchange, electronic funds transfer at point-of-sale, on-line database services, and file transfer applications, are expected to stimulate growth in the use of networks. Network access configurations will vary and 'intelligent' or software-based functionality will be located in the centre and at the periphery of networks.

The intelligent public switched telecommunication network of the 1990s and future decades will rely on an all-digital transmission and switching infrastructure incorporating software, message control and routing databases, and a growing range of data, video, and voice services.[5] This intelligent substructure will accommodate peripheral terminals with their own information-processing capabilities. These will include the telephone set, the calling line identification box, computers and modems, CD-i players, and many other kinds of yet to be marketed equipment. The interworking of disparate network substructure components is expected finally to extinguish any remaining need for the distinctions that now prevail among different types of communication networks. Historically, telephone services, broadcasting, and cable services have been provided using networks that are both technically and organizationally distinct to a greater or lesser degree depending on the country and its legislative and regulatory treatment of these sectors. The catch-all term—multimedia—encompasses a vast range of service applications that is likely to be provided in a broadband network environment. The intelligent capabilities of these networks will support interactive communication between one computer and another, between users and computers, and between users and users. Some of the software features will enable end-user firms or intermediaries, such as systems integrator firms, to alter the configuration of services; others will be controlled by the network operators. Some of the access points to these information highways will be open to the public while others will be open only to designated user groups.

This technical progression of the physical substructure will provide the foundation for increasingly widespread electronic commerce. It has been described by Calhoun as a laminar network.

Despite our historical tendency to think in terms of singular solutions propagated across the entire network—the ultimate version of 'seamless' communications—many observers are coming to the recognition that the most likely future is what is here dubbed the *Laminar Network*: a series of access 'fabrics' overlaying one another, some on a small scale (e.g. microcell radio), some of global proportions [very small aperture terminals for satellite communication], some optimized for narrowband transmissions, others for broadband, some for vehicular communications, others for fixed or portable, some for data-dominant traffic, others for voice-dominant traffic, and still others for the transmission of image-based traffic—all rather imperfectly stitched together. (Calhoun, 1992: 527)

It is in the imperfections of the 'stitching together' of networks that degrees of freedom exist for suppliers and users to influence the trajectories of both substructure and services, and the extent to which electronic commerce strengthens the competitiveness of firms that are able to take advantage of the electronic information environment. Firms that are unable to do so are likely to see their competitive prospects severely diminished in the long term.

The architectural design envisaged for the intelligent public network is contributing to the erosion of historically defined boundaries between domestic and international networks, between local and long distance networks, and between the 'public' and 'private' or corporate networks which characterized the Plain Old Telephone Service era. It is also bringing to an end distinctions between networks which support commercial applications for businesses and consumers, and those which support scientific and technical research and educational applications.

Policy debates on the future of information highways focus on the implications of the convergence of two visions of network and service development which are advocated by the new and traditional operators, respectively. The first has been described as an Internet-based vision; and the second, as an entertainment-based vision (National Research Council 1994).

The Internet vision

The Internet vision is one of an open network that supports any-to-any communication. Internet hosts and users are estimated to number anywhere from 2.5 to 32 million and are spread unevenly world-wide.[6] The Internet uses the underlying communication network substructure to support text, data, and, increasingly, video services. The costs are financed by agencies of the State, that is, in the past by military, and today, by educational and scientific and technical research budgetary allocations, and by contributions from the private sector in support of experimental applications by firms. In the majority of countries, end-users do not pay a fee for accessing the Internet to use services such as on-line databases and bulletin boards, or file transfer and electronic mail facilities.

Many of those who envisage a vast expansion in the capacity and ubiquity of the network substructure also hope that this will yield, finally, an open, permeable, democratic electronic environment. How-ever, this optimistic expectation needs to be assessed against the factors that have contributed to the Internet's development trajectory. The Internet in the USA grew out of a network that was intended originally only for Department of Defense use with its contractors (Hart *et al.* 1992).[7] The private sector always has been involved in the development of the Internet to some extent. For example, the current network was constructed by a consortium including IBM and MCI in the USA,[8] and in other countries there has been private sector involvement especially where telecommunication operating companies have been privatized and where computing companies have contributed to the establishment of high-performance computing test-beds, etc. In the USA, MCI maintained the network for a period of time and IBM provided network management software and switches. In 1994, the National Science Foundation selected four of the Regional Bell Operating Companies to take over the operation of the Internet backbone network in a move that foreshadowed the further commercialization of the operation of the Internet itself, both in the USA and world-wide.

The promotion of the National Information Infrastructure by the US government as well as by private sector companies encourages a vision of a seamless broadband network carrying voice, data, and video services to everyone—including businesses, residential customers, hospitals, schools, and publicly funded research communities. However, publicly financed initiatives to develop the Internet and other networks so that all schools, libraries, and local government offices can be linked into networks supporting information services have met with swift opposition from the private sector. In 1993, for example, AT&T, MCI, and Sprint, three of the main long-distance telecommunication companies in the USA, and the Regional Bell Operating Companies which operate networks in local and regional markets in the USA, stressed that the government should not become involved in financing the network substructure.[9] Similarly, in Europe, the private sector has argued that 'the creation of the information society in Europe should be entrusted to the private sector and to market forces' and this applies especially to the network substructure (Group of Prominent Persons 1994). The public sector role in most countries is likely to be limited to funding research and experimentation, promoting standardization, and protecting the public interest in the privacy of information. At least at the substructure level, therefore, it is difficult to find support for any straightforward transition to the teledemocracy that is championed by some advocates of the Internet vision. It, like its historical substructural predecessors, is subject to the economics and politics of network development.

The entertainment vision

While the Internet vision has so far been championed by those with little or no investment in the network substructure, including computing and software companies, the entertainment vision is being championed by the existing telecommunication network operators and the cable companies. It is also promoted by those companies with major stakes in traditional information-producing industries, that is, the film producers, broadcasters, and established publishing houses. These are all firms that have profited historically from the production and use of information for business or consumer use.

The entertainment vision focuses mainly on residential consumers and the consumption of services that can be provided using existing or modified telephone and cable networks. Services that are at the trial stage include video-on-demand as well as a range of home shopping and game services that have been slowly gaining in market penetration over the past decade. In 1993, in the USA, cable subscriptions generated revenues of some $US 13 bn., film box office receipts amounted to $US5 bn., and video-cassette sales and rentals to between $US 13 and 18 bn. (United States 1994). Whether consumers will dispose of a significantly greater proportion of their income in order to benefit from the new services that are planned is very uncertain. At the same time, the business services market for electronic information services was valued at almost $US 14 bn. in 1993 and this included only services for financial management, research, marketing, purchasing, and general business administration (United States 1994). This figure does not include revenues generated by telecommunication traffic, revenues associated with data processing, or the revenues that would be generated if traffic on the Internet was to be charged on a usage basis. The growth potential of the business information market is likely to create strong incentives for information providers, thus drawing investors away from the development of significantly new and attractive consumer-orientated services.

In Europe as well, in spite of the rhetoric surrounding the need to develop services for the residential consumer, it is business require-ments and electronic commerce that are driving the push to construct the network substructure. The goal is to 'promote the competitiveness of the Community's telecommunications industry, operators and service providers in order to make available to final users the services which will sustain the competitiveness of the European economy and contribute to maintaining and creating employment in Europe' (Commission of the European Communities 1993*a*). Advanced information and communica-tion services are regarded as a source of competitive advantage for firms. The principal reason for the importance of the electronic information services is related to the way such services can alter the timeliness, quality, and diversity of information involved in the conduct of business.[10]

Comparisons of the strengths and weaknesses of the electronic information services markets in the European Union, the USA, and Japan are difficult to make but available evidence suggests that the USA has a clear lead over Europe and Japan.[11] The value of the market for these services is difficult to estimate, in part, because there is little consensus on the definition of the boundaries of the sector. Of particular interest from the point of view of electronic commerce are services which fall within the general categories of audiotex, videotex, on-line databases, CD-ROM, electronic publishing, and multimedia.[12] The sector also includes electronic mail, fax-based services, electronic data interchange, and point-of-sale services. Casting the net wider still, the sector includes services offered by telecommunication and computer network operators as well as those provided via over-the-air broad-casters and cable operators. The business and professional user sectors are the main drivers of the growth of electronic information and communication services. Television and computer-based services targeted at residential consumers are likely to represent a market with much growth potential, but, as in the USA, the price-sensitivity of consumers is largely untested at least on a ubiquitous scale.

The biases of convergence

The network substructure development scenario within the European Union differs little in practice from the USA notwithstanding the rhetoric surrounding the National Information Infrastructure in the latter case. There are examples of co-operation between the public and private sectors in both regions, but this occurs when it is in the commercial interests of the network operators and customers who can be expected to *pay* for services.

Both the Internet and the entertainment visions of the development of the network substructure and services, in principle, can support the widespread diffusion of electronic commerce. Although commercial opportunities for the provision of entertainment are likely to grow considerably there is already a vast market for commercial services supporting electronic commerce. In addition, the Internet, which so far has avoided establishing relationships between costs and prices that are transparent to end-users, is confronting the exigencies of incremental commercialization. Increasingly, businesses are establishing commercial services on the Internet and many of these services support the information elements of electronic commerce. The convergence of these two network visions that policy-makers hope to see could ensure that access to the intelligent network substructure is equitable, affordable, and ubiquitous. There is an expectation that commercial exigencies will not impinge on the behaviour patterns and investment strategies of Internet substructure or service suppliers and users. However, the requirements of electronic commerce are likely to influence both the technical design and operational characteristics of the substructure network as the two visions converge to support the demands of electronic commerce.

Although the Internet model could become the predominant network substructure architectural design because of its emphasis on decentralization, the priorities of electronic commerce will constrain how access and use of this network are managed. Issues such as security, data protection, standards, and the enforcement of intellectual property rights, will create the need for commercial and regulatory controls. The open Internet vision, that is, a vision of a network substructure that is effectively 'free' to all users, would require substantial public investment as demand increases and this is not positively sanctioned by governments in any of the major trading regions of the world. Network closure and proprietary technical designs to achieve competitiveness in an electronic commerce environment will introduce biases towards proprietary network configurations and these design decisions are likely to run counter to the requirements of open networks. The entertainment vision similarly is unlikely to provide a major impetus for the design of the network substructure and services. Demand is more easily determined for professional and business services and this will tend to attract investors towards investment trajectories in these areas in order to reduce risks associated with the launch of new services.

One characteristic common to all these services and networks is their existing or potential interactive nature. Advanced service interactivity extends beyond the transmission of information content, to the manipulation or processing of information. Interactivity can refer to the relationship between the originator of information and the information supplier who packages information as an intermediary between the originator and the end-user; to the relationship between the information supplier and any number of intermediate and end-users; or to the entire chain of relationships. This chain of relationships enables producers and users to create, manipulate, and use information in complex new ways. The organization of this chain of relationships, together with the design of the network substructure, are critical factors in the degree of openness or closure of networks of electronic commerce and how this affects the competitiveness of firms.

REGULATION AND POLICY FOR ELECTRONIC COMMERCE

Firms engaging in electronic commerce also must take into account the characteristics of the regulatory and policy environment. In this section, we highlight some of the salient issues that affect the selection of networking options. The treatment of intellectual property rights in electronic information products is regarded by many firms as perhaps the most significant 'wild card' in the development of new service applications. Insufficient returns to producers of information services which support electronic commerce are likely, they argue, to be disincentives to investment in services for uncertain markets. The protection of intellectual property rights in information is biased as a result in the direction of enforcing tighter controls over the unauthorized use of electronic information services. This issue affects the design of the network substructure and services through its impact on standards for network interfaces and protocols, and software-based applications. Tendencies toward the monopolization of information through stronger intellectual property-rights protection, in turn, creates pressures for the closure of electronic networks, an issue to which we return in Chapter 7. A balance between the interests of the producers and users of information products is inevitably difficult to achieve. This is especially so in an environment where information security and problems in developing encryption devices make it increasingly difficult to monitor uses (and abuses) of existing and proposed legal forms of protection.

The protection of information and the movement of electronic data concerning individuals also are subject to legislation in many countries. Firms engaging in electronic commerce must design their services so that they do not contravene measures to protect individual privacy and these vary substantially from one country to another. Companies in the insurance sector and those engaged in the direct mail advertising businesses, for example, are particularly affected, and many of these firms are concerned about the increasing legal and regulatory bureaucracy associated with the commercial and non-commercial uses of electronic information.

Electronic information services are being offered using networks that cross national boundaries and transborder data flows are growing in magnitude with the increasing mobility of labour and the expansion of international trade. Transborder data flows using all kinds of network substructure are open to the risk of loss of security and privacy if they can be interrupted, intercepted, or modified. The security and privacy of data are therefore major considerations for firms using electronic commerce and this contributes another incentive for network closure.

Digitalization and the extension of advanced network substructures are creating new opportunities for competitive entry in the supply of network substructure and services. National regulations that prevent firms in the telecommunication and cable industries from supplying infrastructure and services in each others' markets are regarded by some companies and public policy-makers as being outmoded as a result of network convergence. It is also argued that, because of the multiple conventional and electronic modes in which information can be accessed, there is no longer any need for restrictions on cross-media ownership restrictions, for example, on newspaper and broadcast station ownership. From one perspective, the relaxation of cross-ownership restrictions can be regarded as a way of creating incentives for the further convergence of the, still segmented, electronic information services industry and the timing of the removal of such restrictions is under discussion in all the major trading regions of the world. Others argue, however, that the market power conferred upon the dominant players by their positions in the markets for information services and network substructure promotes recurring tendencies toward the monopolization of these markets. As a result they advocate the retention, and in some cases, even the strengthening of existing measures to protect new entrants and consumers from anti-competitive business practices.

Generally, the firms engaged in the design of electronic commerce networks have relatively parochial views on the extent to which national regulations and legislation influence their selection of network and service configurations. They argue simply that there is a need for transparent and coherent regulations and policies to promote the production and use of electronic commerce. The need for the regulation of cross-ownership of different media, the regulation of dominant infrastructure network operators, and for restrictions on foreign ownership, are perceived differently depending on the market positions of firms and whether they offer their views with regard to activities in their home markets or in the distant markets they seek to enter. Large firm size, market dominance, and longevity in the market tend to be associated with the view that anti-competitive practices legislation in the home market should be relaxed and that policies to attract foreign-owned companies should be complemented by reciprocity. Small firm size and recent entry into the network substructure or information services market, in contrast, tends to be associated with the view that legislation to protect firms from the market power of dominant firms must be enforced (Mansell and Tang 1994).

These domains of electronic commerce—the design of the network substructure, of services, and of the regulatory and policy environment—are intertwined with the organizational features of electronic commerce. In the next section, we look at the factors that lead to network differentiation as a result of incentives in the organizational domain. Organizational factors are commonly addressed in studies of innovations within firms and their use of advanced information and communication technologies. However, it is the negotiation of outcomes in all four domains that is critical to the incentives for the design of open and closed networks. These outcomes are fundamental to the design of electronic commerce, its biases, and its implications for the public interest in the ubiquity and affordability of new generations of networks and services.

THE DESIGN OF ELECTRONIC COMMERCE SYSTEMS

The growth in the use of electronic commerce reflects a shift which has been documented by many observers (see Lamberton 1984; Machlup 1962; Office of Technology Assessment 1994), from the production and consumption of goods to the production and consumption of services. However, shifting commerce itself into the electronic sphere does not necessarily bring markets closer to the perfection imagined by neo-classical theorists. Instead, it multiplies the number of parameters that firms must take into account in their strategic behaviour. In so doing, it creates a more complex environment for firms seeking to enhance their efficiency and competitiveness by breaking down constraints imposed by time and geographical space (Harvey 1989).

We have considered three of the four dimensions of the selection environment that firms face. In most studies of the development and use of advanced information and communication networks and services by firms, these dimensions are simply the backdrop or 'environment' for analysis of intra- or inter-firm electronic networking activities. However, electronic commerce results from a combination of interactions among numerous kinds of institutions in each of the four dimensions. Together, these create a 'system of innovation' which governs how firms organize the conduct of business in electronic space. As Niosi *et al.* (1993: 218–19) have pointed out, 'it is not one type of institution that creates a system, but the simultaneous existence and pattern of interaction of a series of institutions, and the feedback that they receive from the surrounding environment.'

The diffusion of advanced information and communication technologies and numerous institutional changes in substructure, services, policies, and regulations, are preconditions for electronic commerce, but they are not sufficient conditions. Firms also initiate electronic commerce in the light of their internal organizational capabilities. Here, we need to consider a significant number of factors. Case-studies of the design and implementation of electronic trading networks are used here to illustrate these and to suggest that, while there is significant latitude for firms to shape the components of the network substructure and services to meet their needs over time, there are also indications that internal organizational factors create strong biases toward network closure rather than toward the open vision which characterized so much of the policy rhetoric in the 1990s.

Electronic trading networks represent a subset of electronic commerce networks. They enable all or some of the information elements involved in market exchange to be carried out electronically (see Figure 4.1 above). The emergence of electronic trading networks is associated with two developments which occurred in the mid-1980s. The first was the use by large firms of networks to establish electronic markets to support financial and related services. The second was the growth of the intelligent public telecommunication network on an international scale. The decision to include trade in services in the Uruguay Round of the GATT negotiations in 1986 reflected the significance of these and related developments in the global use of new combinations of computing, software, and telecommunication equipment. In the early 1980s, network market-places were described in the following terms:

Network information services connect the needs and resources of users to the capabilities and services of producers and facilitate transactions between them. All the usual services of a marketplace can be offered within a large information network. . . . Indeed, the delivery of products and services for business, industry, the consumer, and government can be perceived as a marketplace in the emerging information society—a marketplace on a communications network or the *network marketplace*. (Dordick 1981: 1)

Electronic communication networks not only were contributing to the tradability of information services, but traditional physical markets themselves were becoming increasingly immaterial. Little was known about the factors that would encourage firms to participate in electronic commerce or that would contribute to the design of proprietary (exclusive) versus common (open) standards for the interactive exchange of information.

If the development of electronic trading networks is to conform to the 'open vision' described in the preceding section, it will need to incorporate technical and other measures to ensure that biases or distortions in the market favoured by firms and governments are, at the very least, transparent, and at best, eliminated. However, the interests of firms in using electronic networks to gain sustainable competitive advantage leads them to 'design in' network features that encourage network closure. In these circumstances, incentives to implement open network interfaces are reduced and incentives come into play which encourage the adoption of proprietary standards, encryption techniques, and other practices that secure network environments from intrusions by competitors.

Our case-studies of electronic trading networks focused on users, operators of networks and services, and the producers of the hardware and software components of networks. The design features and operational characteristics of networks and services were examined and the constraints and degrees of freedom available to firms to initiate creative responses to potential conflicts of interest among the actors involved in network design were also considered. Case-studies of seven electronic trading networks in the Netherlands, Sweden, the UK, and the USA provided the basis for the analysis of the factors contributing to network design. The user firms included large and small firms and the case-studies were drawn from the financial services, distribution and transport, and agriculture and horticulture sectors. They included:

1. A market operation for the cross-border trading of stocks run by a company based in the UK owned by a Swedish holding company. In order to make the electronic trading network secure, proprietary standards were adopted and the system was effectively closed to users other than large institutional investors.

2. A global after-hours system for 24-hour global trading of financial instruments jointly launched by British and US investors which adopted proprietary standards and measures to reduce risk to medium-sized and multinational investors using the trading system.

3. A scheme initiated in the UK to introduce an electronic system for the reconciliation of oil and gas products among large firms. Although the scheme was successful in devising standards (electronic data interchange standards) that were taken up by the European Commission, it was not implemented on-line due to potential threats to the security of information.

4. A scheme to support the activities of a port authority in the UK which adopted proprietary standards to protect the network from competitive systems. It was closed to users located at other port authorities due to user concerns about competitor's access to information.

5. A Dutch tele-auction providing an electronic market for the fruit and vegetable trade. The network users regarded the service as a way of locking non-Dutch products out of the market and it was implemented using proprietary standards in a closed network environment.

6. A Dutch network supporting the auctioning of flowers and pot plants. Users regarded the network as a vehicle to strengthen their role as dominant players in the co-ordination of cross-border trade but were unable to expand the capacity of the system to meet their needs because of restrictions in the regulatory domain.

7. A cargo information system developed for users mainly in developing countries using public funds. This electronic trading network was designed in co-operation with public authorities and was open to a multi-user community. Concerns were expressed about the viability of the design once private sector operators took responsibility for operating the system.

In all of the case-studies, it was the operators of the technical substructure or of the information services who forged new institutional relationships among user firms. In cases (1) to (7), when the network operators' interests coincided with those of the members or users of the network, the networks underwent changes and were upgraded from a technical point of view, but they remained essentially closed to users outside the initial communities of traders in the various sectors.

Among the organizational features of the firms that influenced their preferences regarding the design of these electronic trading networks were size, strategy with regard to collaboration and competition in core markets, the locus of control ranging from significant autonomy to direct or indirect control by parent organizations, and ease of access to capital. Users of these electronic trading networks also were substantially influenced in their selection of network designs by the degree to which they had in-house expertise or needed to acquire capabilities externally to implement and use the electronic trading systems. Technical choices also were influenced by the steepness of the learning curves associated with different aspects of network implementation and maintenance.

Conflicts were experienced between user firms and the designers of the networks in all of the case-studies, not because of the absence of viable technical solutions, but as a result of difficulties in gaining agreement on sensitive issues regarding the accessibility of networks and the security of information. The technical problems in each of the case-studies were resolved through a continuous process of learning and experimentation on the part of the network designers and users. The main problems encountered were the uneven availability of appropriate network substructures, the costs and access conditions of the network substructure and services in different geographical areas; the need for co-ordination across spheres of competence among public telecommunication operators, cable operators, third-party systems integrators, software developers and users; and, finally, the need to agree technical standards that would not jeopardize the integrity or security of the information and communication services.

CONCLUSION

In some theoretical models electronic trading networks are expected to enable market-related information to be exchanged without regard to geographical constraints of place and time. They are expected to reduce transaction costs and to bring the functioning of markets closer to the ideal of perfect competition. However, in our case-studies, all the actors involved had found ways to articulate their economic interests through the technical and organizational design of the networks.

The firms and organizations participating in these examples of electronic trading were not primarily restricted in their choice of network design by geography or by technology, but rather by the need to cluster the advantages of regulations and policies, electronic services and substructure with their distinctive and changing capabilities and competitive strategies.

Each of the case-studies suggested that a high degree of coherence had been forged among network suppliers and users and that this was maintained through time by drawing on a mix of internal and external resources including capital and skills. In most cases, however, firms turned to external sources of capabilities only when they could foresee no perceived threat to the security and integrity of the information they produced and used in connection with their core business and trading activities.

The case-studies also suggested that the operation of electronic trading networks tends to be controlled and governed within those national jurisdictions which offer the most favourable regulatory conditions. This applied to regulations affecting network substructure and services, and to those affecting financial restrictions on cross-border trading.

Harold Seidman argues that 'the quest for co-ordination is in many respects the twentieth century equivalent of the medieval search for the philosopher's stone. If only we can find the right formula for co-ordination, we can reconcile the irreconcilable, harmonize compelling and wholly divergent interests, overcome irrationalities in our government structure and make hard policy choices to which no one will dissent' (Seidman 1980: 205). Electronic trading networks, and electronic commerce more generally, are expected to enable collaboration among competing firms and, in theory, the elimination (or reduction) of the information asymmetries which contribute to market imperfections. Differences in the economic and political interests of the actors are expected to be reconciled such that open network configurations are favoured by suppliers and users. Our case-studies suggest, however, that this is not the predominant trend. Rather than regarding electronic commerce as a precursor to unbiased markets, it is useful to consider the extent to which the biases in electronic markets lead to advantages or disadvantages for firms that find themselves either included or excluded from participation.

The design features of the commercial world underlie the convergence of computing, telecommunication, and audio-visual technologies and it is these features, together with the way they are coupled and uncoupled by supplier and user firms, that affect the openness and closure of the underlying technical networks.

As Kapor argues, most important are:

the *design principles* (both in technical architecture and public policy) under which the winning entry or entries operate . . . Openness is either present or absent in every aspect of the network. A network is either open or closed with respect to who may have access to it, who determines its specific uses, who may supply content, who can provide the equipment used, how interfaces and standards are determined, and whether the technical details are public or private. (Kapor 1993: 57)

The determinants of innovation in electronic commerce are to be found in the specific contexts that govern the selection and implementation of changing technical systems (Dosi and Orsenigo 1988). Different local or national contexts create varying incentives for the development of electronic commerce. These differences are the keys to the relative openness or closure of the electronic network substructures and services that will emerge in different countries. As one of the best illustrations of the 'borderless' world envisaged by authors such as Kenichi Ohmae (1990), electronic commerce networks might be expected to show signs of neutralizing the advantages of historically accumulated economic and political power. Instead, new technical configurations and network designs are being institutionalized in ways that reflect disparities in firms' capabilities to innovate and to compete in world markets. Thus, technical innovations in advanced information and communication technologies are being reconfigured in the light of existing patterns of control, they show few signs of eradicating disparities in firms' capabilities, and they show tendencies toward network closure although this is often expressed in innovative ways.

There is, therefore, a continuing role for public policy that seeks to redress disparities in order that the public interest in open networks can be articulated and, in appropriate contexts, implemented. The public interest in the design of advanced information and communication networks is taken up in the following chapters through examination of the implications of advanced networks and services for consumer surveillance, the standardization process, and the international governance of these complex technical systems.

NOTES

1. See Rothwell (1992) for the results of work on the characteristics of the '5th Generation Innovation' process and Mansell (1995) for a review of literature on the implementation of information and communication technologies within and between firms. Bloomfield *et al.* (1994), Fleck (1994), Kimble and McLoughlin (1994), and Liff and Scarborough (1994) offer a range of case-studies drawn from experience in the UK. The Massachusetts Institute of

Technology 'Management in 1990s' programme produced case-studies of the experiences of firms and their strategies toward the implementation of information and communication technology networks, see Scott-Morton (1991) for a collection on the MIT programme and Taylor and Van Every (1993) for similar work in the Canadian context. See also, Commission of the European Communities (1990) for case-studies in the European context.

2. An exception to this observation is the 'markets-as-networks' research tradition in Sweden at Uppsala University and the Stockholm School of Economics, see Johanson and Mattsson (1994) for a review.

3. For earlier versions of this matrix see Mansell *et al.* (1991), Mansell and Jenkins 1992*a*, *b*, 1993).

4. The perfect competitive market is one in which all participants have equal access to information, and, in which, at equilibrium, all sales are executed at a market-clearing price and there are zero transaction costs (Hirshleiffer and Riley 1979). Departures from the perfect market are attributed to the costs of searching for information (Stigler 1961) and to variable non-zero transaction costs (Coase 1937).

5. This network includes a Service Control Architecture with information processing elements; a Service Management System which runs on a computer and updates Service Control Points with new data or application programs and collects statistics about the performance of the network. Service Control Points differentiate between basic and newer services. Signal Transfer Points are part of a Common Channel Signaling System No. 7 (CCSS7) network, a standardized communication interface which links the network's intelligence to the switches embedded in the rest of the physical network. The Service Switching Point is the point of network access for all users of the network.

6. Internet access nodes were located in 55 countries in 1993; 25 of these were high-income countries; 12, upper-middle-income countries; 12, lower-middle-income countries; and 1 (India), a low-income country using UN definitions (International Telecommunication Union 1994). These nodes connect with domestic academic, scientific, and technical research networks within countries. Estimates of the number of hosts and users vary depending on the way they are calculated. Estimates of the number of hosts in July 1994 ranged from 707,000 to 3.2 million, and of users from 2.5 to 32 million (Hoffman and Novak 1994).

7. From 1969 to 1983 ARPANET expanded to over 100 nodes but access was limited to defence agencies and contractors and the network operated at a maximum speed of 56 kbit/s. Special purpose networks were built in the 1980s on the ARPANET model by the National Science Foundation, university grants, and the private sector. The Internet protocol—TCP(Transmission Control Protocol)/IP(Internet Protocol)—is less efficient for some applications than more recent developments but became a *de facto* standard incorporated into early versions of the AT&T UNIX system and all UNIX systems now contain the TCP/IP kernel. IBM and DEC now support the TCP/IP interconnection services as supplements to their proprietary network protocols.

8. MERIT, a private sector company in the USA, was awarded $US 14 m. by

the public sector to put the network in place. Industry contributions to the National Science Foundation supported network, NSFNET, to 1989 by MCI and IBM are estimated at $US 40 m. to $US 50 m. as compared to the NSF $US 14 m. contribution (Marshal 1989).

9. The House of Representatives received a proposal from Congressman R. Boucher (Chair of the subcommittee on Science, Space and Technology) to revise the 1991 High Performance Computer Communications Act.

10. See esp., David and Foray (1994*a*), Jonscher (1983), Lamberton *et al.* (1986), Melody (1981, 1987), Monk (1993), and Moore (1991).

11. See European Commission (1994), Mansell and Tang (1994), and United States (1994) for reviews of the difficulties encountered in comparative analysis in this sector and an overview of the USA, European, and Japanese markets.

12. The US market for all electronic information including consumer services was estimated at ECU 10.5 bn. in 1993. The European Union market for professional users of publicly available electronic information services was valued at ECU 4.2 bn. in 1992 of which on-line services accounted for ECU 3.6 bn. (European Commission 1994).

5

Surveillance by Design:
Public Networks and the Control
of Consumption

ROHAN SAMARAJIVA

INTRODUCTION

As human interactions increasingly shift from face-to-face, proximate, and synchronous forms, to electronically mediated, distant and/or asynchronous forms, of interaction it is not surprising that elements of control found in all social relations are becoming apparent in electronically mediated interactions. What is noteworthy is that many electronically mediated interactions are being subject to stronger forms of control. Increasingly, the very medium of electronic interaction is being designed as a pervasive surveillance instrument serving control purposes.

Control and surveillance as used in this chapter carry no pejorative connotations.[1] Following Beniger (1986: 7–8), I define control as increasing the probability of a desired outcome rather than its absolute determination. Giddens (1981: 63) emphasizes the dialectic of control arguing that 'however wide-ranging the control which actors have over others, the weak nevertheless always have some capabilities of turning resources back against the strong.' All actors have options although the scope and intensity of control they find in a changing environment varies in scope and intensity (Giddens 1987).

In this chapter, I am concerned with control as extended over time and space. Control emerges in many different ways in the modern State and Mulgan (1991), for example, has distinguished between conceptions of exogenous and endogenous control. Exogenous control is understood to be that which is imposed, abstracted, and rationalized, while endogenous control refers to that which is communicative and shared. The former is associated with tools, weapons, techniques, and struc-

The valuable criticisms of previous drafts by Robin Mansell, Patrick Hadley, Huichuan Liu, and Chris Richter are gratefully acknowledged. The discussion of space in this chapter was developed in collaboration with Peter Shields.

tures; and the latter with language, commonalty, self-regulation, and nurture.

Giddens (1979: 206) situates control within time and space through his concept of locale which he argues is a preferable term to 'place'—'for it carries something of the connotation of space used as a *setting* for interaction'. Locales enter into social reproduction by creating and sustaining the taken-for-granted meanings of everyday routines. Giddens suggests that space is a resource or context drawn upon in interactions but he provides little insight into how various kinds of spaces are produced (Gregory 1989). For Giddens, locales are primarily defined in terms of proximity, and co-presence and electronic media are primarily instruments that facilitate 'time-space distanciation', or the integration of distant locales. Left unexplored are the virtual spaces produced when electronic media are used to 'bridge' or 'annihilate' physical distance.[2]

Giddens (1987) introduces surveillance as a technology that takes the form of the accumulation of coded information utilized to administer the activities of subjects, and the direct supervision of the activities of persons in authority. He suggests that direct supervision by organizations is limited to segments of subjects' lives and, primarily, to that portion spent at work in an office or factory. I argue, however, that direct supervision is being extended beyond the office and factory to 'leisure time'.

Capitalism is necessarily technologically dynamic and fraught with uncertainty and insecurity. As Harvey (1989: 105–6), following Marx and Schumpeter, argues, there is 'perpetual flux in consumer wants, tastes, and needs' and this is a permanent site of uncertainty and struggle. There has been a shift from consumption of goods to the consumption of services which, through their ephemeral nature, allows capitalists to transcend the limitations to accumulation and turnover of physical goods that have characterized earlier periods (Harvey 1989: 285). Bressand *et al.* (1989: 17) go further to argue that the trend is toward merging goods and services into *compaks* which they describe as 'complex packages in which goods coexist with services'. Bressand *et al.* (1989) challenge the analytical primacy commonly accorded to production over consumption, pointing to the increasing role of symbols in consumption.[3]

Although those such as Lyon (1994: 157) argue that consumption, being based on pleasure, is fundamentally non-coercive, control in the sphere of consumption and control in the political-legal sphere belongs to the same continuum.[4] In the control of consumption what is involved is the manipulation and creation of preferences, life-styles or packages of interconnected preferences, and narratives which the agent adopts (Jhally 1987). The agent experiences pleasure as well as continuous dissonance and insecurity. The sphere of consumption and the role of surveillance, particularly by information and communication technologies, is only beginning to be considered in the literature (see Flaherty 1985; Gandy 1993; Lyon 1994; Robins and Webster 1988).

My analysis in this chapter utilizes control, surveillance, consumption, and associated concepts to identify the structural forces responsible for the form of the design of the increasingly pervasive virtual spaces wherein humans interact with each other and engage in transactions. I discuss the conduct of key actors who are contributing to the emergence of trends that are yielding public virtual spaces conducive to the control of consumption and sketch out possible scenarios for future development.[5]

The phenomenon under investigation is perhaps most developed in the USA and Canada, but even in these countries, the existing public communication networks are rudimentary in relation, as has been shown in Chapter 4, to what is possible with existing technical capabilities. In most other parts of the world, the redesign of networks is lagging behind North America and the study of concrete efforts to redesign the electronic communication network provides a sound basis for analysis of emerging trends.

VIRTUAL PUBLIC SPACE[6]

A locale, as defined by Giddens (1984: 13), is a power container wherein actors generate/concentrate allocative and authoritative resources.[7] Actors draw upon a common awareness of properties of the setting of interaction or locale to achieve meaningful communication. In this way locales enter into social reproduction by creating and sustaining the taken-for-granted meanings of everyday routines. The 'production' of locale is not a neutral activity (Gottdiener 1986; Soja 1989; Weisman 1992). I propose the use of a cluster of related socio-spatial concepts—space, environment, public space, private space, proximate space, and virtual space—to understand the biases in the processes of design that are at work in the particular case of public communication networks.[8] Space is understood as a terrain of social interaction and it is produced, reproduced, and transformed by the same structural forces, social relations, and conflicts which affect social life more generally (Lefebvre 1991). As a social product, space is a material force that reflects back upon social processes, a process described by Soja (1980, 1989) as the socio-spatial dialectic.

The simplest form of proximate space is constituted by the co-presence of human actors who are knowledgeable of each others' existences. Aspects of an environment, for example, the existence of walls or high noise levels, play an important role in the constitution of a proximate space. The potential for communication, for instance, is crucial, because communication, broadly defined, between the actors is a precondition for the existence of a social relation which, in turn, is the precondition for the existence of a social space.

The design of physical environments enables the exercise of power over actors. In his analysis of shopping malls, the modern exemplar of the designed environment, Gottdiener (1986: 293) argues that the function of mall design is 'the control of crowds to facilitate consumption . . . to disguise the exchange relation between producer and consumer . . . and to present cognitively an integrated façade which facilitates this instrumental purpose by the stimulation of consumer fantasies.' The specific relations, or lack thereof, between actors in an environment determine the nature of the proximate spaces that can be created within it. These relations are shaped, in turn, by factors such as the comparative location of actors in social structure, their perceptions of these locations, and their perceptions of previous interactions and expectations concerning future relations. Depending on the nature of relations between co-present actors in a given environment, multiple proximate spaces can be created in one physical environment. A prisoner occupying a proximate space with his guard is constrained from creating a new space with a third party, for example.

A complex form of proximate space is produced when actors have asymmetric knowledge of each other. The space constituted by a voyeur utilizing the communicative properties of a one-way mirror is an example. The 'big brother' mode of surveillance associated with Orwell (1954) takes place in this form of space. A more complex spatial form is constituted when people behave as though others are potentially co-present. The production of this space is central to the disciplinary power postulated by Foucault (1977). Foucault uses the panopticon, an architectural device advocated by Bentham in 1791, to exemplify the process giving rise to the self-policing subject. The panoptic space enables a guard to see the prisoners, but to be invisible to them. Since prisoners cannot be certain when they are being observed, they must assume constant surveillance: 'He who is subjected to a field of visibility, and who knows it, assumes responsibility for the constraints of power; he inscribes the power relation in which he simultaneously plays both roles' (Foucault 1977: 202–3).

In terms of my conceptual framework, the subject is 'locked in' to a coercive space with a power-wielder, that is, subjects watch their own behaviour so they can do what they think is the power-wielder's will. Liberation from this space comes only when the absence of surveillance is established. Foucault does not allow for this. In his vision, subjects are never 'alone'.

Until the advent of electronic media, most social interactions occurred in proximate space, but virtual space is rapidly emerging as a significant terrain of human interaction. In virtual space, actors achieve rudimentary forms of co-presence via electronic communication media ranging from two-way video to the much more ubiquitous telephone. Virtual spaces must be constituted from electronic environments and physical environments.

Being relatively more malleable than physical environments, electronic environments are more conducive to dynamic and continuous exercise of control through the technical features of networks and pricing practices. Electronic environments can be designed to enable pervasive and transparent surveillance through the tracking of usage patterns and long-term storage of such information.

Both proximate and virtual spaces may be divided into public and private spaces. Goffman (1963: 9) has defined public spaces as 'any regions of a community freely accessible to members of that community'. A more precise definition would begin from the concept of public environment. Public environments are those not designated as private by permanent or temporary markers or signs. Highways, sidewalks, plazas, and public parks are physical examples. Public, as well as private, spaces can be created from public environments. The range of possible spaces extends from the strongly private, for example, lovers on a park bench, to the strongly public, for example, a police officer directing traffic. Most of the possible spaces occupy intermediate points on the continuum, for instance, eye-contact and body signals between pedestrians navigating a busy sidewalk. Public spaces are characterized by relative openness to initiation of communication by others, and private spaces are characterized by relative closure to initiation of communication. Negotiation of the terms and conditions of communication is an intrinsic and continuing element of every communicative relationship (Petronio 1991). While public spaces may be conceptually distinguished from public environments, in practice, the two are interchangeable because, in many cases, other individuals and their behaviours provide the markers of 'publicness'.

Virtual public spaces are constituted from electronic and physical environments. The public communication networks and computer networks of the mid-1990s constitute virtual public spaces.[9] These networks enable the initiation of dyadic or small-group communication with others among millions of connected individuals, and one or more among these millions initiating communication with oneself. The 'publicness' of a space depends on openness to initiation of communication among inhabitants rather than on whether anyone can come into the space. The existence of subscription fees for accessing a network does not, therefore, vitiate the publicness of a virtual space.

The moment of 'entering' virtual public space by lifting the telephone handset or logging on to a computer network is similar to entering a proximate public space. Contact may be initiated with a person or persons in that space at that moment, for example, a chatline or the 'talk' mode in the Internet; or later when the person or persons log on; or by initiating contact with those in physical private and public environments abutting the virtual public space, such as a telephone subscriber at home or in a pub with a telephone.

Generally, actors do not establish interpersonal contact with totally unknown persons in virtual or proximate public space.[10] It is more common for individuals to navigate public space to establish contact with a known person or persons, at which point the dyad or larger group effects a complete or partial withdrawal from the public space into a private space. In both kinds of public space, the possibility of unintentional collision exists in the forms of physically bumping into bystanders and dialling wrong numbers.

In both proximate and virtual public spaces, boundaries between public and private spaces are defined by implicit, and sometimes explicit, negotiation between the communicating parties, but also between them and third parties. Infringements of these boundaries or the use of coercion in the negotiating process constitute violations of a basic ground rule identified by Goffman (1971: 198):

[I]n Western society, as probably in all others, there is the 'right and duty of partial display'. Two or more individuals present together have the right and duty to make some information generally available concerning their relationship and the right and duty to leave unsignaled other information about their relationship.

Drawing on Goffman's work and subsequent scholarship (see Altman 1975; Petronio 1991; Ruggles 1993), privacy is defined here as the capability to implicitly or explicitly negotiate boundary conditions of social relations. This definition includes control of the outflow of information that may be of strategic or aesthetic value to the person and control of the inflow of information, including initiation of contact.

The proposition that the public communication network and electronic mail networks are virtual public spaces may be challenged by the apparent lack of temporal co-presence between the inhabitants of the virtual public space. It may be difficult to accept the claim that the millions of potential called parties are co-present on the network at any given moment. They may be out boating, at lunch, or dead. The frequency of 'telephone tag' may be used to refute the claim of co-presence. Nevertheless, the claim is defensible in so far as social space is constituted when one party believes the other is potentially co-present. Discovery of the other's absence terminates a private space. In a public space, the expectation is not one of certain communication with a particular person, but of potential communication with some person. The absence of a particular individual does not negate the public space.

It is possible to treat humans as existing in proximate and virtual space and to consider their interactions in both kinds of spaces. How power in one kind of space is leveraged into the other; how control over environment is translated into power in the resultant space; and how recursive practices of agents in these spaces produce and reproduce structural power, are issues that are at the core of the redesign of public communication networks and the role they will play in surveillance.

The surveillance that is the focus of this chapter is that which is increasingly being designed into the virtual public space of public communication networks, the relatively ubiquitous, primarily voice or data-only networks, that date back to the nineteenth century. Despite their age and, in the industrialized countries, taken-for-granted nature, these networks are undergoing massive changes in almost all countries of the world (Mansell 1993a; Wellenius 1993). The emerging network configuration will form the basis for the various forms of information infrastructures and 'highways' being promoted across the world (Gore 1994; Huth and Gould 1993) and it is to the career or life-cycle of this trajectory that we need to turn if we are to understand the social and economic implications of the resultant design.

The overall phenomenon of surveillance and control of citizens and consumers is much larger than that which involves public communication networks (Gandy 1993; Larson 1992; Lyon 1994; Mukherjee and Samarajiva 1993). However, the almost 'universal' nature of the connectivity afforded by these networks sets them apart from other modalities and justifies special attention (Samarajiva 1994). While other forms of surveillance in society tend to affect segments of life (Poster 1990), the enhanced public communication network of the future is likely to affect a much larger portion of everyday life. For example, plans for the information infrastructure in the 1990s comprise the delivery of entertainment services including games, banking services including bill payment and transactional services, lotteries, telemetry for energy and other utilities, security services including perimeter and internal motion detection systems, in addition to voice, video, and text communication services including library retrieval. The integration of different records of behaviour that today occurs imperfectly and out of the view of the subject, will tomorrow occur in the infrastructure, or the virtual public space, itself.

SURVEILLANCE IMPERATIVES

The qualitative change underway in the capitalist economy has been characterized by many writers as holding the potential for a shift from mass production to flexible customized production, implying a reorganization of production, distribution, and consumption processes (Aglietta 1979; Castells 1993; Piore and Sabel 1984; Roobeek 1987).[11] Changes in the production sphere have been relatively well documented, but less attention has been paid to changes in the distribution and consumption spheres and their interrelationships. Bauman (1988: 61) suggests that, in the present period, at least in the Western industrialized countries, capitalism takes a form in which a majority of individuals compete in the symbolic terrain of consumption, insulating the production sphere from excessive strain and, indeed, strengthening it through increased demand for products of capitalist industry. The melding of the moments of production, distribution, and consumption in relation to the role of mass marketing and mass media (Smythe 1981: p. xiv), has advanced greatly as a result of the 'informatization' of economic processes.

It is now routine, for example, for companies such as Honda America to switch their assembly lines from one make of car to another, or from production to maintenance or training, based on real-time sales data flowing in from dealerships. Major American consumer products companies such as Toys 'R' Us and The Limited occupy the interface between production and distribution instead of actually engaging in production. Their retail outlets are wired into sophisticated sales reporting networks which analyse the movement of products at an extremely fine-grained level in real time. Based on this information, the head office, which is not necessarily a single, fixed site, modulates the pace and nature of production that actually takes place in far-flung corners of the world at plants that also are connected to the head office via sophisticated communication networks. These plants are not owned by the distribution company, but are contractually bound to it. The distribution company occupies the pivotal position through its control of trade names and associated marketing mechanisms, and its control of the sales reporting network or surveillance capability.

The nature of marketing is also changing. Mass production of standardized goods required mass advertising campaigns to shift individual preferences toward a norm or to create a norm that matched the mass-produced object.[12] This was the basis of mass advertising where identical messages were pumped out to large audiences; and of mass marketing where large numbers of consumers were assembled in department stores and supermarkets. As flexible customized production emerges, it is no longer as necessary to rely on mass distribution of identical messages or on large retail stores.

Increased clutter in conventional mass media outlets, such as network television, drives down the cost–benefit ratio of mass advertising (Peppers and Rogers 1993). As the relative costs of information and communication technologies, such as customized printing presses, telemarketing, and computer-based marketing systems decline, more and more marketing campaigns shift from the mass form to the flexible customized form. As markets become more fragmented, specialized stores in malls and direct marketing become more effective. Consumer surveillance is a precondition for the success of flexible marketing and this, in turn, requires information as to whom the goods, services, or compaks should be customized for.

The flexible production-marketing emerging in the USA is not production to satisfy the given, sovereign needs of consumers discovered through consumer surveillance. There is continuous effort, at a fine-grained level, to discover crucial characteristics of consumers, the levers of their behaviours or their 'hot buttons'. These characteristics form starting-points for two related efforts. The first is the effort to produce persuasive messages that will yield specific purchasing decisions. The moulding of preferences continues, but it is now increasingly more sophisticated and invidious. The second effort is to connect specific subsets of consumers to specific channels of distribution that will enable the successful linking of the supply of goods, services, or compaks, to demand, moulded or modified to correspond to that supply.

Consumer surveillance takes internal and external forms. In the first, the enterprise seeks to convert what used to be anonymous transactions of goods into transactions encompassing goods and services, or compaks. Where some service relationship existed, it seeks to strengthen the relationship and increase its information-generating capabilities (David Shepard Associates 1995).

Whereas automobile dealers were once content to sell cars and take the cash, these high-value transactions are increasingly being converted into service relationships. The car is bundled with warranties and sometimes with twenty-four-hour emergency service and maintenance packages which increase the service component of the transaction and require the yielding of personal information, including access information, to the dealer.[13] The US government's regulations regarding motor vehicle recalls makes it a legal requirement that automobile dealers and manufacturers maintain records of purchasers' addresses. In many cases, manufacturers and dealers are promoting extremely attractive lease arrangements, which convert what once were product transactions into service transactions.

The frequent travel award programmes pioneered by airlines, and now utilized by credit-card companies, gasoline retailers, and even supermarkets in the USA and elsewhere, are classic examples of strengthening the information-generating features of service relationships. These programmes position customers to voluntarily report details of travel including destinations, duration, hotel preferences, and rental car usage, that can then be used for customized marketing (David Shepard Associates 1995), and for shaping the production and marketing processes. The blurring of the producer–consumer distinction postulated by Bressand *et al.* (1989) appears less a cause than an effect of the need to subject consumers to surveillance.

As enterprises learn more about consumers through detailed, continuous, and systematic surveillance of consumers' interactions centred around a compak, the ensuing knowledge is applied to: (i) the production and communication of more effective persuasive messages to more perfectly align the symbolic needs of the consumer and the symbolic elements of the compak; (ii) the development of a smoothly functioning channel for connecting consumer demand for and supply of that particular compak; (iii) the extrapolation of the consumer's symbolic needs from the first compak (known) to other compaks (unknown) through persuasive messages; and (iv) the connection of the consumer's demand for different compaks, created or modified in (iii) above, to supply, via the channels developed in (ii) above. In other words, the alleged freedom of the consumer (Bauman 1988; Lyon 1994) is increasingly reduced; the control exercised by the enterprise over the consumption phase is increased.

Internal consumer surveillance creates the basis for external surveillance of the market in consumer data. The extension of control from one market to another described in (iii) and (iv) above need not be intra-enterprise. Consumer data collected by one enterprise can be purchased by others and utilized to create or modify demand for different compaks. The selling of consumer data or the 'leasing' of transactional communication network channels is becoming an important line of business for enterprises. In some cases, the generation of consumer data can become the main business of an enterprise (David Shepard Associates 1995: 27). External and internal surveillance are not mutually exclusive. According to the conventions of the trade, responses to mailings or calls made to prospects on a rented list belong to the renter (Novek *et al.* 1990). Once prospects are converted to customers, internal surveillance can be activated. Data quality is important for both forms of surveillance. Because a conventional, that is, pre-compaks, relationship between the consumer and the marketer is of the win-lose type, consumers have incentives to dissimulate. The creation of compaks and surveillance relations changes the relationship giving it some win-win characteristics because incomplete or inaccurate data will reduce the quality of service subsequently received. In addition, data supply by consumers may be hindered by transaction costs. As a result, consumer surveillance tends to favour involuntary forms of data supply by consumers that are essentially transparent. This is called Transaction-Generated Information (TGI). If the data are generated as by-products of transactions, their quality is necessarily higher and there are no additional costs to the consumer.

To utilize TGI, data on past behaviour, to influence future behaviour, assumptions and inferences, are necessary. Knowledge in this area is rather crude, for example, such as in cases where it is assumed that inhabitants of a geographical area designated by a postal code share common characteristics (Weiss 1988). But such assumptions will become more and more finely tuned and they can be expected to improve dramatically over the next few years as more data become available, analysis tools become more sophisticated, and more marketing research is done.

Examples of TGI collection and analysis abound. Next to the National Security Agency, the official code-breaking and eavesdropping agency of the US government, American Express is the most important customer for Cray supercomputers. This is because American Express engages in extensive analysis of TGI from the charge-card transactions of its upscale clientele. Many supermarkets in the USA are inducing their customers to use in-house electronic payment cards or identification cards so that customer-specific records of purchases can be maintained (Larson 1992). Inducements range from easier cashing of cheques to customized rewards in the form of discount coupons that are generated when a specific volume of purchases is recorded. The recording of these most mundane of transactions highlights the extent to which the 'threshold of describable individuality' (Foucault 1977: 191) has been lowered.

Any consumer surveillance reduces freedom in the sphere of consumption. Actions by organizations that collect TGI in order to create or modify subsequent consumer behaviour and the narrowing and channelling of consumer options diminish that freedom. Two subsets of TGI deserve particular attention because they are especially deleterious to consumer autonomy. These are communication-related TGI and TGI collected by firms with significant market power, particularly monopolies.

The former category is made up of TGI pertaining to communication interactions and transactions. Examples are records of who called whom, when, and for how long, which is also referred to as Telecommunication Transaction-Generated Information (TTGI), and records of an individual's video or audio use.[14] TTGI gained attention during the Watergate scandal in the USA that toppled President Nixon. Telephone company records revealed that one of the Watergate burglars had placed several calls to the office of the President's re-election committee (Bernstein and Woodward 1974: 35–6). Telephone records also provided evidence that a call had been placed from ice-skater Tonya Harding's home in the USA to the practice site of her opponent, thereby connecting her with the beating of Nancy Kerrigan. While US legislation enacted in November 1994 tightened the previously easy access to TTGI by law-enforcement personnel, on the negative side, telecommunication operators in the USA now are required by law to facilitate wiretaps on their networks (United States Congress 1994).

Entertainment-related TGI gained notoriety when a newspaper *United States* published the video rental records of Judge Robert Bork in the midst of a heated battle over his confirmation as a member of the Supreme Court (Miller 1991: 9). This led to the Video Privacy Protection Act of 1988 prohibiting disclosure of titles of rented movies and the release of individually identifiable rental records in the USA. The sensitivity of interaction and transaction records derives from their enabling nature; to do almost anything one needs to communicate with someone. In a market-based society it is difficult to exist without transactions.

Knowledge of 'who talked to whom and when' can yield valuable inferences even when what was talked about is unknown. Information transaction records are valued because of the common perception that information preferences provide a clearer indication of personality than other indicators, as evidenced by the frequent mention of music or movie preferences in personal advertisements. Coupled with this is the common suspicion that individuals do not disclose all their information preferences, particularly those of a sexual nature.

TGI collected by firms with significant market power, particularly monopolies, is more valuable than general TGI and poses distinct problems. The best examples of this second type of TGI are data collected by companies that provide infrastructure services such as energy, communication, sewage, transportation, and water, under exclusive franchise. In theory, these firms should have few incentives to collect TGI, since their customers have no alternative sources of supply and demand for the service is relatively inelastic. In reality, despite the fact that collecting TGI is not a zero cost activity for the producer, such incentives have emerged as by-products of particular forms of regulation. Where these companies are allowed to engage in competitive and monopoly activities, that is, monopoly service provision along with 'competitive' services, incentives to collect TGI have emerged and pose significant policy problems.

For example, in competitive markets, customers facing intrusive information collection have the option of moving to less privacy-invasive suppliers. This option is unavailable to customers of the monopoly arm of firms operating in mixed monopoly and competitive markets. The transfer of information gathered by the monopoly arm of the firm to the competitive arms skews the 'playing field' in the competitive markets, damaging the prospects of competitors in the short term and customers in the long term, and harming public policy objectives in both instances.

Furthermore, given the vital and enabling functions performed by infrastructure facilities in all aspects of life, these transactions are likely to yield more information about core patterns of customer behaviour, be they other firms or individuals. Whereas TGI collected by a competitive firm is, by definition, limited to a segment of the relevant market and/or the populace and is of limited value in terms of drawing population-wide conclusions about behaviour, the TGI collected by a monopoly firm more or less covers the entire market and is thus more valuable (Burns *et al*. 1992; Mukherjee and Samarajiva 1993).

TGI collection by telecommunication service or cable service providers who have so far retained their monopolies, or a substantial degree of market power, is doubly problematic because these companies are collecting communication-related TGI. Traditionally, advertising-supported mass media firms such as newspaper and broadcasting companies have had little or no formal transactions or ongoing relationships with their readers, listeners, or viewers. It is not that these companies have not been desirous of learning about the behaviour and proclivities of customers. In the past, however, this desire had to be satisfied by explicit information-seeking acts such as audience surveys, ratings, and competitions. The increasing tendency of advertising-supported firms in the USA to enter alliances with telecommunication companies and outright mergers, and to provide services with inter-active elements such as pay-per-use weather and sports services, and audiotex services tied to classified personal advertisements, suggests that consumer surveillance is likely to spread through the entirety of the information and communication network and services sector.

The 'information superhighway' rhetoric in the USA, Canada, the European Union, and elsewhere, which gained currency from 1993 onwards suggests that all consumers will be connected through one or two conduits to an integrated broadband network constituted by the convergence of the public switched telecommunication network with the cable network and the Internet, yielding a virtual public space that will be an integrated network with varying degrees of regionalization. The TTGI produced by interactions and transactions in this virtual public space will be extremely valuable for the control of consumption. Significant conflict is likely, therefore, over the locus of control of TTGI.

Due to its relative ubiquity, and interactivity, the existing public communication network can serve as a prototype for the future network or 'network of networks' (see Noam 1994*b*). Indeed, conflict over TTGI generated by interactions and transactions has already erupted in the USA and some other countries, the most prominent example being calling line identification (Samarajiva and Shields 1992).[15]

I argue that these conflicts cannot be resolved simply by passing legislation on specific information collection and dissemination practices, though such action may be useful within specific contexts. If the communication network is redesigned to incorporate surveillance as an integral element of virtual public space, the degrees of freedom available to agents who acquire different capabilities to subsequently change it will be highly constrained.

A key question is, therefore, whether the public communication network is being redesigned to incorporate surveillance. Establishing that incentives for surveillance exist does not necessarily lead to the conclusion that surveillance is being built into the network by design. Mansell (1993a) establishes that the network is being redesigned to allow public telecommunication operators to offer new services in a flexible manner. She makes a convincing case that the redesign reflects specific choices often favouring these operators against their competitors. Out-of-band signalling, or the transmission of call set-up information outside the frequency band used for communication by the user, is identified as the key technical innovation which creates flexibility in the public communication network (Mansell 1993a: 26). The technology being installed in networks world-wide for out-of-band signalling is Common Channel Signalling System 7 (CCSS7). Mansell's discussion emphasizes the competitive implications of the evolving network redesign, but does not address the user surveillance aspects.

In early North American applications for approval of 'no-block' calling line identification service (where the calling party was not allowed to prevent transmission of the calling number), telephone companies claimed that 'providing a blocking function [with calling line identification], as some advocate, would significantly reduce, if not totally negate, the benefits that Caller ID [calling line identification] has proven it can deliver to telephone users' (Ohio Bell Inc. 1991). Underlying this claim is the argument that since a person placing a call cannot know with certainty whether the called party has the number display capability, he or she will have to assume it. This will make the calling party behave 'appropriately'. The companies argued that blocking, if provided, would be selected by the criminally deviant, for example, obscene callers. In a no-block environment, all callers are likely to behave with the expectation that call-receiving parties can receive and record their numbers and calling times. This argument has significant parallels with the logic of the panopticon. It does not matter whether an individual does not subscribe to calling line identification; it does not matter if the people that the individual calls do not subscribe to calling line identification; as long as the possibility exists that someone subscribes, the individual will self-police him or herself.

No-block calling line identification service is not unusual, nor is it an aberration from the general pattern of new telecommunication service offerings. The 'intelligent' network is being designed on the premiss that the calling party's number can be recognized and utilized (Gnoffo 1990). When combined with personal telephone numbers, the information generated by the future public communication network will be of a different order of magnitude as compared to the present. Further foreshadowings of surveillance in the network are offered by ongoing efforts to detect fraudulent telephone use. Cellular network operators and business telephone users are installing software in their systems that will trigger alarms upon noticing 'deviant' calling patterns (Adelson 1993; Lewyn 1992). The corporate adjustment to the rising importance of network surveillance is indicated by the trend toward organizing customer accounts by telephone numbers.

What in particular is worrisome about such a network redesign? The public communication network is a public space and any diminution of autonomy in public space is cause for concern. The network is now more like a village green or a common, with the associated freedoms and dangers. One usually walks across the village green to meet previously known persons and create private spaces of interaction, but the green also allows for the initiation of contact with previously unknown persons, creating conditions conducive for buying, selling, and persuading. The green allows for solitude in company, a paradoxical but nevertheless familiar and pleasurable experience. The green allows for surreptitious trysts as well as the possibility that such trysts will be discovered. It is not an entirely peaceful place, having the potential for serious bodily harm in the form of muggings and rape. The public communication network is similar. It is usually used to create private spaces with known others, but it can allow for serendipitous meetings. It is a potentially dangerous place, allowing for obscene and harassing calls which, in most countries, are criminal acts. Interestingly, it is impossible to detect or deter an obscene or harassing caller who does not make repeated calls under the 'old' design of the network.

Designing surveillance into this virtual public space is akin to posting large numbers of police on a village green. Safety, if defined as protection from deviant members of society, may be increased. When the dangers of police coercion arising from the problematic definition of deviance (Cohen 1985) are factored in, even the gain in safety is questionable. The fundamental issue, however, is the loss of serendipity intrinsic to public space because of surveillance.

The problem with the public communication network redesign runs deeper than mere observation of comings and goings. The surveillance that is proposed is of a qualitatively different kind, such as that found in high-security buildings. In high-security buildings the doors 'remember'. As an individual passes through he or she must yield identification information by swiping a card or providing voice or retinal prints. This TGI is stored, and may be analysed for patterns and deviance on an as-needed basis or routinely. After enough time passes, the collection of TGI will become transparent for most users, and continuous surveillance will be taken for granted. Even if the data are not analysed regularly, users will act as though they are, as long as ambiguity and uncertainty are maintained.

High-security private networks which are virtual private spaces do not have negative implications *per se*. By definition, their private nature allows for restrictions and features different from public spaces. The primary concerns are raised by their effect on access to public space (Noam 1991) and their impact on other important social policy objectives such as protection of children, and prevention of unfair labour practices (Samarajiva 1996 forthcoming). However, the hardwiring of pervasive surveillance into virtual public spaces does violate privacy and is therefore harmful *per se*.

The surveillance being built into public communication networks is not panoptical in a pure sense, although it appears that its designers have panoptical vision. The efficacy of the panopticon depends on the inmates being knowledgeable about the gaze and fearing the consequences of the detection of deviance. Research has shown, however, that few people have knowledge of the increasingly pervasive commercial surveillance to which they are subject (Gandy 1993). Furthermore, periodic spectacles of punishment are necessary to create fear of punishment (Samarajiva and Shields 1992: 410). Whether the panoptic vision will be realized fully, depends on the introduction of mechanisms for creating and reinforcing knowledge of the gaze, through periodic spectacles of punishment or otherwise. The 'drug war' and offensives against computer hackers and cybercriminals may provide opportunities for 'educating' the public and completing the design. Concern over the safety of children of the type that has led to a rapid increase in closed-circuit television monitoring in Britain offers another possibility.

In the USA, 'toll-free' telephone, or 'freephone' services as they are more generally called in Europe, have become common and provide a mechanism to collect TTGI prior to the emergence of the full-fledged intelligent public communication network. Transactions of all sorts are increasingly completed through such services. Direct-mail or catalogue companies, such as L. L. Bean in Maine, now receive more than 80 per cent of their orders via freephone services. The commercial uses of these services in the USA extend far beyond direct mail. Purchasers of Nintendo computer game devices, for example, are provided with an 800 toll-free or freephone number at purchase time. This number can be used to obtain 'over-the-shoulder' advice from Nintendo technicians on how to connect the device to a specific television monitor. Mundane, low-value consumer products, such as pancake mixes and toothpaste, carry 800 numbers on their packaging. It is becoming more difficult to reach organizations such as airlines and some government offices without utilizing freephone telephone services.

The design of toll-free or freephone services in the USA and elsewhere incorporates surveillance. In the early days, the calling party numbers were recorded and periodically provided to the subscriber as part of the invoice. Since the 1988 approval of AT&T's INFO-2 tariff, calling party numbers have been transmitted in real-time to INFO-2 subscribers. The calling number of the incoming call is, in many cases, used to route the customer record associated with the number to the screen of the customer service agent prior to the call being answered (Carnavale and Lopez 1989). In some cases, the numbers are cross-matched with reverse telephone directory databases to generate mailing lists for solicitations (Levy 1991), or are sold as lists of telephone numbers to telemarketers. In all cases, the telephone numbers are captured and stored by the subscriber company. In the UK, and the European Union more generally, data protection legislation and national telecommunication regulatory authorities are seeking to ensure that TTGI is not available for reuse by third parties but it is unclear as yet whether such safeguards are enforceable.

Surveillance is becoming a routine feature of the service offered by telephone companies around the world. It is surveillance of a rather crude form in that most actions triggered by the recording of the number assume calls originating from a home or office telephone and that the calling party is the subscriber of record for the call-originating telephone. The surveillance is generally transparent to the persons who make toll-free or freephone calls and, generally, has been uncontroversial (Federal Communications Commission 1994).

However, another form of network-based surveillance places even greater weight on TTGI. This is the development of customer profiles by telephone companies based on calling patterns. For example, in the USA, AT&T markets international calling plan services based on analysis of an international caller's call destinations, frequency, and other related TTGI. Some companies offer calling pattern and billing analysis as a service to their customers (O'Shea 1993), but generally the customer's permission is not requested. Other instances of TTGI utilization include development of marketing profiles for sale of enhanced services and customer premises or terminal equipment by Regional Bell Operating Companies and the target marketing to frequent toll-free or freephone users of directories of toll-free services by AT&T (Miller 1991). In these cases, individuals receiving the marketing solicitations or the directories may suspect some sort of surveillance, but will be unable to pinpoint the precise source of information because the targeting is based on a set of their telecommunication behaviours. The customer will be unable to withhold access to the data since they are routinely generated as by-products of his or her telecommunication usage.

AGENCY IN THE RISE OF SURVEILLANCE BY DESIGN

The customer's capacity to influence the design process of elements of the intelligent network, and, particularly, the software package that includes the calling line identification capability, has been examined by Samarajiva and Shields (1993). Understanding of this capacity or agency on the part of customers requires elucidation of the manner in which the structural forces of competition and regulation are translated into the design of the software. This, in turn, requires participant observation and interviews with relevant actors to obtain information on how the actual designers of the public communication network located in R&D laboratories became 'knowledgeable' of the structures as they designed the software. The skills and other dimensions of the software development process have been discussed in Chapter 3 and in this chapter I focus directly on the attributes of this process that are specific to design of software embedded within the public communication network.

'Knowledgeability' refers to 'everything which actors know (believe) about the circumstances of their action and that of others, drawn upon in the production and reproduction of that action, including tacit as well as discursively available knowledge' (Giddens 1984: 375). Everything includes agents' motives, beliefs, and intentions as well as their 'knowledge of the strategic terrain which faces or faced an agent and which constituted the range of possibilities and limits to the possible' (Stones 1991: 676). The information-intensive characteristics of most advanced information and communication services suggest that these structural factors have been translated into network design features, but that this process has not been complete. For example, the flexibility built into the software package for calling line identification services, such as the ability of a customer to prevent the delivery of calling party numbers, suggests that the translation has not been absolute.

Clearly, technical capabilities that may be designed into a software-based communication service do not constitute the totality of a service or determine the way a service will be implemented and used. Marketing and regulatory processes also affect the final shape of the service. The outcome of the agency of the technical designers 'congeals' in the software and constitutes part of the structure for the marketing and regulatory personnel. Nevertheless, marketing personnel have a degree of freedom in their sphere as well. In the USA, Canada, and the UK, marketing and regulatory considerations have led to suppression of some of the technical capabilities of the software package for services such as calling line identification and to the accentuation of others.

Pricing and the bundling of software functions to create services have reflected decisions that shape the services that are offered for regulatory approval and consumer consideration.[16] For example, one telephone company, Ohio Bell in the USA, originally offered only two services, calling line identification with no blocking capabilities, and a simple form of Call Return, choosing to suppress most of the capabilities of the package and all the number-delivery-blocking features (Ohio Bell Inc. 1991).

As a participant in the regulatory process in the USA and Canada wherein the calling line identification service, as defined by the telephone company marketing and regulatory affairs personnel, was approved with modifications, I am well informed about how the technical capabilities and the symbolic representation of the service constrained and enabled policy-makers and other stakeholders. There was plenty of room for choice at the regulatory stage, as evidenced by the widely divergent outcomes of the calling line identification service decisions by the 46 state regulatory commissions that took up the issue in the USA (Mukherjee 1994). Our analysis suggests that frames for thinking about calling line identification changed over the period since the service was originally proposed (Mukherjee 1994; Samarajiva and Shields 1992; Viswanath *et al.* 1994). These frames were structures produced and reproduced by agents in the policy process.

The telecommunication regulatory process in most countries is reactive, usually being activated by the formal filing of a tariff by a telephone company. As a result, policy actors often have to act within the bounds of the technological features that are presented to them after the software development process has 'congealed'. In the case of calling line identification, the technical capabilities of the software package gave them more room to act because of the 'bundled' nature of the software package and the various options built into it during the technical design phase. For example, the efforts of Ohio Bell marketing and regulatory personnel to have only two services considered, proved unsuccessful. Advocates of public interest concerns successfully petitioned the regulatory body to require the company to offer non-privacy-invasive services such as Call Trace, a service that competed with calling line identification in helping customers deal with obscene and harassing calls, and introducing the suppressed number-blocking features of the calling line identification and call return services into the policy debate.

Had these features not been included at the technical design stage, the only choices open to those advocating consideration of broader public interest concerns would have been to accept or reject the services as described in the tariff. Instead, they were able to fashion an electronic environment different from that proposed by the telephone companies by 'mixing and matching' the technical features of the telephone companies' own software.

The end result of the above design process has differed in various jurisdictions in the USA and in other countries, but it is becoming part of the electronic environment of users of the public communication network. For example, users in New Jersey, the first American state to approve calling line identification, and one of the few to allow no number-blocking options to residential users at the outset, faced an electronic environment that was quite different from that of adjacent Pennsylvania, where the highest appeals court held the service to be in violation of state wire-tap laws, and hence illegal. In New Jersey, users who know about the service and its capabilities have to factor in the likelihood that their numbers could be displayed to the called party before the telephone is picked up and possibly stored and further processed. In Pennsylvania, there was no change in the electronic environment.[17] In the European Union it seems likely that the implementation of the complex software that supports calling line identification and related services will be implemented with variations among the member States.

Agents' knowledgeability of characteristics of the electronic environment shapes the kind of virtual spaces that are produced. If calling parties are unaware of this new capability of the network or of the possibility of its utilization, their behaviour will not be affected. In most American jurisdictions, less than 10 per cent of users subscribe to this service. In early trials in 1994 conducted by British Telecom in the UK less than one per cent of customers used the 'call-blocking' option and only one customer in 5,000 requested 'line-blocking'.

Knowledge that the called party lacks the equipment to receive the number could yield an entirely different outcome. It is possible, for example, to envisage an intimate, such as a spouse or lover, telephonically harassing a partner even in a no-block environment based on the knowledge that the victim has no calling number display device. Thus, an environment which leaves the fewest options for users does not completely 'determine' what users actually can do. Based on knowledge of the structural features of the network, users could act in various ways that are not directly predictable from the environmental features. Nevertheless, their actions are constrained and enabled by structures, including the congealed structure of the architecture of the network and actual behaviour patterns can only be discovered from the results of user studies (Lievrouw and Finn 1990; Shields *et al.* 1993; Silverstone and Morley 1990).

These illustrations suggest that the interplay of structure and agency yields an indeterminacy, within a certain range, that should gladden the heart of critics of privacy and surveillance writings who decry the passivity induced by dystopian analyses based on the Orwellian metaphor of 'Big Brother' and the metaphor of the panopticon (Lyon 1994). The public communication network operators may have the incentive and desire to redesign the network as a panopticon, but, absent the meshing of actions by multiple actors, the outcome cannot be certain. If the populace does not know that the panopticon is in place, or if a significant proportion actively subvert it, the panopticon will be imperfect. This is not to suggest that opposition to the panoptical design process is unnecessary. Indeed, oppositional actions are among the factors giving rise to indeterminacy in the process.

ALTERNATIVE NETWORK SURVEILLANCE SCENARIOS

As communication network providers in the USA and other countries are gradually freed from their traditional public service obligations, they undergo corresponding organizational and cultural changes. In this environment, public communication operators are likely to respond favourably to the demand for high-quality TGI from business customers and, in addition, to generate such information for internal use as the operators become increasingly market orientated. The addition of a strong real-time surveillance component to toll-free or freephone services utilized by large organizations illustrates the process. The calling line identification service is an extension of surveillance capabilities throughout the network. The panoptical vision underlying this redesign, if fully realized, would transform the principal virtual public spaces of industrialized societies from ones characterized by relatively weak control, to ones with much stronger control features.

Whether or not this design will be fully realized cannot be definitively answered here but it is possible to outline the features of a number of likely scenarios. In an *Integrated Network I* scenario, public telephone network and service providers enter into alliances with cable companies, value-added or specialized telecommunication carriers and cellular companies, and absorb the Internet, recreating, more or less, the integrated electronic environment that existed historically in the USA and elsewhere, but at a qualitatively higher level. Control would be relatively centralized and TTGI would flow freely through the network. While there would be other companies interconnected to the public network, they would be subservient to the principal operating entity.

If the operators of the integrated network stay out of advanced information and communication service activities, they may earn the trust of companies doing business over the network and be allowed to perform surveillance and market facilitation functions for them. In this scenario, the resistance is likely to come from consumers whose privacy and autonomy would be considerably weakened. Efforts to go 'off-network', introduce anonymous payment systems, and similar responses are likely. However, without significant support from information and communication content and service providers, resistance is unlikely to be very effective.

Should the operators of the integrated public communication network decide to participate fully in information and communication content and value-adding activities, which is the present tendency, the equation is likely to be significantly different. In the *Integrated Network II* scenario, they will be perceived as real or potential competitors by a multitude of other content and service providers where such competitors have been allowed to spring up. These competing companies will have incentives to develop network-independent market-facilitation mechanisms as have computer bulletin-board service providers who use credit cards for billing and collection. This scenario is friendlier to encryption, electronic cash, and similar privacy-enhancing technologies, than the preceding one. Arrayed against the public network operators will be a phalanx of service providers and consumers. In this scenario, the likelihood of legislation curtailing strong surveillance would be stronger.

A third scenario, the *Fragmented Network*, could emerge from the inherent instability of the *Integrated Network II* scenario. Lack of trust in public network operators on the part of service providers as well as consumers could create the conditions for a collapse of the integrated network as more and more 'off-network' functional service attributes are utilized to substitute for, or subvert, public network design features. An example is the increasing reliance by many service providers, including long-distance operators, on independent billing and collection services such as credit-card companies, despite the fact that public communication networks in most industrialized countries are superbly equipped to fulfil that function. This scenario is heavily dependent on advances in, and legal permission for, cryptographic applications such as electronic cash.

The foregoing scenarios assume that a supply-push, producer-dominated marketing process is at work. The closest analogy is the home-shopping network. In the home-shopping services, the marketers advertise their wares and employ hosts to sell products and services. Viewers who tune in to the channel use touchtone telephones, toll-free or freephone services, and credit cards to complete transactions. The purchased item is delivered by a parcel delivery service. Advanced versions of home-shopping services, being promoted by the likes of Barry Diller, former Chief Executive Officer of the Fox Network in the USA, include hypertext capabilities built into the programming, whereby viewers interested in particular compaks featured in the programme 'leave' the programme and enter advertising or trans-actional virtual spaces created by the marketers. However, interactive, the network designs engendered by the supply-push model privilege exogenous control.

The alternative is a demand-pull model where consumers have more power. Transactions would still occur on the network, but they would be initiated by consumers rather than by producers. Control would cease to be exogenous. While this is a radical break from the present trend toward a deepening of control in the consumption sphere, it is closely related to developments in the flexible production economy. Many of the changes in the production sphere have occurred in producer–producer relations, particularly in the relations between companies that assemble final products and their suppliers. Lean production techniques such as just-in-time require suppliers to supply only on-demand. A demand-pull model would extend this principle to the consumption sphere. Marketers would be able to advertise their wares only when consumers wish to enter the market. The closest analogy would be to the classified, printed 'yellow pages' that many consumers peruse when they wish to purchase specific compaks. In the demand-pull model the incentives for customer surveillance and control would be replaced by incentives to allow demand to be demonstrated efficiently. Production would respond to demand rather than seeking to direct it. Here, the public communication network would become a truly competitive market-place, facilitating the effective flow of accurate and desired information between suppliers and demanders and enabling the efficient conduct of transactions.

The demand-pull model may develop in different ways. First, in the case of a *Centralized Market-Facilitating Network*, a centralized, integrated network such as described in the *Integrated Network I* scenario, could offer market-facilitating services within a demand-pull frame-work. In order to earn the trust of service providers and consumers, operators would have to stay out of content and service activities.

For example, the public network operators could allow customers to open time-limited 'windows' with specified product or service para-meters whereby marketers could provide large amounts of information about their products or services to consumers who have specifically indicated their interest. After the 'window' is closed, marketers would not be able to pursue the customers. Information outside the specified parameters would not be accepted.

In this scenario, the public network operators would accumulate large amounts of sensitive TTGI. Therefore, they would require extensive privacy and security safeguards. The acceptance of their services being dependent on trust, they would have an incentive to show that the sensitive TTGI is not misused, and that it is not vulnerable to attack from third parties. Strong security safeguards, security and privacy audits, and overall regulation would have to be part of this scenario.

Second, in the case of a *Decentralized Market-Facilitating Network* scenario, there would be multiple networks, not all of them fully integrated but allowing some forms of interconnection. Some of the dangers of centralized data collection discussed above could be minimized by creating the conditions whereby TTGI is fragmented. There would also be many potential providers of competitive networks.

Some of these, such as utility and rail companies, have 'deep pockets' and rights of way. For example, as energy companies install fibre to optimize their residential and business energy networks using telemetry, there are incentives to provide telecommunication services using the excess capacity. If the public network operator continues on its high-surveillance trajectory, it is possible, though not certain, that the other network operators would seek to distinguish themselves by offering safeguards from excessive surveillance.

These scenarios do not exhaust the possibilities, but they illuminate the multiple potential trajectories that are embedded in public communication network development. In these scenarios, I have emphasized the potential role of corporate actors, particularly in shifting the public telecommunication companies from their current trajectory of designing a virtual public space that enhances surveillance. Greater weight has been given to the commercial aspects of the future communication network than to its non-commercial aspects. It is unlikely that activism on the part of public interest groups by itself will be an effective counterweight to the structural forces favouring the design of a high-surveillance virtual public space.[18]

CONCLUSION

Some authors posit the existence of a sphere of consumption that is free of control (see Bauman 1988; Lyon 1994). I have shown that the control imperatives existing in the sphere of production are also at work in the sphere of consumption. This is a result of the integration of production, distribution, and consumption, and of the extension of 'ever more potent and costly technological means of gathering and producing information' (Bauman 1988: 59) from the production and distribution spheres to the consumption sphere. It may be that the degree of control is less significant in the consumption sphere than in the production sphere. However, my analysis of leading-edge developments indicates that control of consumption utilizing information and communication technologies is a process that is well underway. I also suggest that information and communication technology-intensive surveillance pervades all aspects of control, in contrast to Giddens's (1987: 13–15) claim that surveillance is associated solely with the generation of authoritative resources.

My examination of the redesign of virtual public space in the context of the public communication network suggests that, unlike the private domains in which mechanisms of consumer surveillance are being installed, these public spaces afford greater possibilities for intervention by citizens and public advocacy groups. The common and potentially conflictual use of virtual public spaces by multiple organizational actors leaves open relatively larger possibilities for effective policy intervention to address public interest concerns.

The primary reasons for focusing on public communication networks are linked to the normative values associated with public space and privacy. In all societies, citizens attach great value to freedoms associated with public spaces, despite the dangers and risks associated with such spaces. It may be argued that the serendipity and openness to initiation of communication characteristic of public space is essential to a society's dynamism and growth. Along the same lines, it may be suggested that privacy is an important and valued element of communication in all societies, though it takes different forms in different cultures. 'Hardwiring' virtual public spaces for surveillance is a fundamental assault on both these normative values. The present direction of public communication network redesign is likely to face increasing resistance as these threats to fundamental normative values become clear.

Moral outrage, however, is unlikely to change the trajectory of powerful structural forces. Understanding the potentialities embedded in the existing structures and identifying those agents who are likely to give rise to alternatives is a key task of research in the critical political economy tradition. Effective communication of this knowledge to policy actors can increase the probability of changing the existing trajectory of network redesign from one favouring exogenous control to one favouring endogenous control.

It is likely that the main impetus for alternative trajectories will come from conflict between traditional public communication network operators, on the one hand, and competing information and communication content or service providers and organized consumers on the other, or, indeed, from rivalry among competing network infrastructure providers. As long as communication network operators pursue entry into these markets, their relations with content and service providers will be fraught with tension. This tension is likely to create the conditions for the emergence of privacy-enhancing mechanisms such as electronic cash, and for the offering of different functional service attributes by competing network and service suppliers. Combined with informed public policy and the public's, non-volatile, yet growing concern about privacy, contradictions between the actions of corporate actors may yield less dystopian virtual public spaces than those initially suggested by current trends. The functional service attributes embedded in the public communication network are reflected in technical standards and protocols whose careers or life-cycles also are open to considerable uncertainty and change. The degrees of freedom available to agents in the standardization environment are the subject of the analysis in the next chapter.

NOTES

1. Giddens (1981: 50) refutes claims that the related terms 'domination' and 'power' are inherently noxious and coercive and this argument is extended here to 'control' and 'surveillance'.
2. Giddens emphasizes the substantially given character of 'space'. He gives less emphasis to what Gottdiener (1986), Harvey (1973), Lefebvre (1991), and others call the 'production of space'. Giddens's concept of locale also does not distinguish between proximate and virtual space.
3. On the increasing importance of symbols, see e.g. Bauman (1988), Bourdieu (1984), and Giddens (1991).
4. Bauman (1988), who is drawn on extensively by Lyon (1994), suggests that consumption is marked by freedom or, implicitly, lack of coercion, but in Bauman *et al.* (1992: 142) the nature of this freedom is described as 'a "velvet repression" . . . free consumers . . . gladly, willingly, joyfully enter the dependency relation with marketing companies, with experts, technological or scientific.'
5. The analysis is grounded in a critical realist methodology (Bhaskar 1989; Sayers 1992) and influenced strongly by Giddens's (1984) theory of structuration.
6. This section is adapted from Samarajiva and Shields (1993). In previous writings, the terms 'electronic' and 'physical' were used instead of 'virtual' and 'proximate'. The former was confusing because space is defined as a relational term, and electronic and physical are not relational. The previous terms are retained as adjectives for 'environment' which is not a relationally defined term.
7. Allocative resources are defined as 'dominion over material facilities, including material goods and natural forces that may be harnessed in their production' and authoritative resources as 'the means of dominion over the activities of human beings themselves' (Giddens 1987: 7).

8. These concepts are drawn from Giddens and a burgeoning multidisciplinary literature which includes Benedikt (1992) and Gibson (1984, 1987, 1988) on the concept of cyberspace.

9. Reference to the public communication network refers mainly to the public switched telecommunication network but is not limited to switched networks of this kind.

10. Actors engaged in political or commercial activity wish to initiate contact with previously unknown persons. Allowing access to unknown persons is one of the functions of a public space.

11. Piore and Sable (1984) claim that flexible specialization may replace mass production. Reich (1992) claims that it will. This author's position is that mass production is in crisis, that it is undergoing fundamental transformations not all of which have been analysed conclusively. While mass production still dominates in volume, the trend is away from it. Mass production can be made to appear to be customized, for example, individually addressed and 'personalized' letters with the same content produced by high-speed, high-volume presses. In addition, the continuing fragmentation of media audiences is ending mass marketing, necessarily affecting the outcome of changes in the production and distribution sphere.

12. The moulding of consumer preferences is fraught with a fundamental contradiction. On the one hand, preferences are being moved toward a common norm. On the other, each consumer must feel that he or she is unique and doing better than his or her peers. Veblen (1899) was the first to illuminate this fundamental feature.

13. Gardner (1988: 384) defines access information as 'information disclosed in public that can be used to locate an individual at some future time. Knowledge of a person's home or place of work clearly can function as access information, as can knowledge of one's full name, phone number, neighbourhood, habitual routes, and hangouts.'

14. McManus (1990) first proposed the term 'Telephone Transaction-Generated Information'. With an information infrastructure that combines current telecommunication, broadcasting, cable, and computer networks the scope of TTGI will be very much broader.

15. Other controversies pertaining to control of TTGI in the USA are documented in McManus (1990) and Burns *et al.* (1992).

16. Additional services such as call waiting and calling line identification are generally marketed in sets (bundled), where customers are given price incentives to purchase more than one service in a set. Services in a bundle have related technical features, or are portrayed as related to each other in the marketing effort. Calling line identification was seen by the telephone companies as the 'anchor' service of the bundle of services introduced in the latter half of the 1980s in North America and a similar approach is being adopted in European countries. It is unusual in requiring the purchase or lease of a display or storage device. The software package that includes calling line identification includes several services such as call reject, call trace, and various forms of number delivery blocking. No significant cost savings could be realized by not offering a subset of the services included in the package.

17. The design of the electronic environment showed a degree of fluidity over the period of study. In 1987, New Jersey became the first to approve calling line identification. In the original regulatory ruling, no forms of blocking were allowed. In 1993, following a petition by the American Civil Liberties Union, the telephone company agreed to provide free per-call blocking for the general public and per-line blocking for women's shelters. In late 1993, the Pennsylvania legislature amended its wire-tap statute to enable the provision of calling line identification with per-call and per-line blocking. The efforts of the Federal Communications Commission to regulate this service gained momentum in 1994, though all the issues were not fully resolved by the end of the year (Mukherjee 1994).

18. This is despite the high level of concern about privacy among the public at least in North America (EKOS Research Associates 1993; Louis Harris and Associates 1993).

6

Standards for Communication Technologies: Negotiating Institutional Biases in Network Design

RICHARD HAWKINS

INTRODUCTION

Communication systems cannot function without standards. Even the most trivial face-to-face exchange is difficult or impossible except in the presence of linguistic codes and semiotic points of reference that are mutually available to the communicating parties and mutually understood by them. Communication as a human act has been mediated by technology since at least the invention of writing. As a consequence, it also has been linked closely to technical standards. Alphabets, codes, systems of numbers, units of measurement, and so forth are basic communication technologies, but they are of little use unless they are standardized to a degree such that they can be applied and understood amongst a collectivity of potential communicators.

To varying extents, the designers of electronically mediated communication systems strive to reproduce or emulate many of the conventions of interpersonal communication. At the same time, they add many new conventions to the repertory of communication practices. Analogous to interpersonal communication, the various technological components in an electronic communication system also require commonly recognized codes and references in order that information can pass from one component to another.

This chapter addresses the subject of technology standards for the interconnection and interoperation of electronic communication systems. The primary emphasis is on *tele*communication networks, but many of the issues raised are common throughout the entire spectrum of information and communication technologies. In much the same way that a standard alphabet is a cultural artefact as well as a utilitarian device, standards for electronic communication networks are products of social, political, and economic relationships as well as being critical technical components. In their absence, these networks simply could not function. This general axiom holds, irrespective of the specific form of a standard, or the specific context in which it is applied.

The primary concern is not with the role of standards in conferring degrees of functionality on networks—this is an engineering problem. Rather, the concern is to examine the central relationship between the technical necessities for standards and the many non-technical criteria that enter into the standards selection process. Standards can (and do) embody many political and commercial biases and still perform more or less adequately as technical elements in the network. The problem for network designers and operators is to ensure that the standards employed do in fact deliver the required levels of technical functionality.

The fundamental question for social scientists is: 'Given the critical role of standards in providing for operational functionality in electronic communication networks, how are the technical and non-technical criteria synthesized in the process of selecting and applying standards?' This question is highly relevant in establishing criteria for the evaluation and application of standards, and it is also important in policy terms. Although standards-making is not a random process, it is fraught with uncertainties and difficult to control. If policy-makers fail to understand the special dynamics of standards and standards-making, the result can be unreasonable expectations, and questionable allocations of resources.

The theme of this chapter is that the selection, application, and evolution of technical standards in the communication industries is influenced by structural alignments and realignments among the social, economic, political, and technical institutions that interact in the overall design of electronic communication systems. Standards relate to both the functional and aesthetic aspects of design as described in Chapter 1. In functional terms, they relate to the structure and configuration of the physical components of a technologically mediated communication system. In aesthetic terms, they relate to the *appearance* of the system to the communicating parties, in terms of the access conditions, the ease or difficulty of use, and the flexibility of the system in meeting changing needs.

The following discussion draws upon the principal findings from research carried out by the author over approximately a four-year period. The foundation for this research was an immersion study of the standardization process in the telecommunication sector. This involved detailed examination of primary documentation from standards organizations, and of the published standards themselves. It also incorporated extensive fieldwork involving in-depth interviews with officials in industry, government, and standards organizations, and a protracted period of participant observation in some of the technical committees in which standards are developed.

The immersion study yielded many insights into the kinds of

relationships that exist between technical and non-technical factors in standards-making, and, more importantly, into the actual dynamics of the process—the ways in which these various factors wax and wane in influence over time, and how the resulting standards reflect specific kinds of technical and non-technical influences. These insights were then applied in further thematic studies. The first study looked at events in the European Union, and questioned the practice of using voluntary standards as instruments for pursuing public policy goals. The second examined the role of business users of telecommunication services in standards-making. The third investigated the evolving roles of voluntary standards in defining the relationship between public and private networks, with a concentration upon developments in mobile telephony. The final thematic study focused on the place of standards in the design of the global telecommunication infrastructure.

The first part of this chapter is a discussion of definitional, operational, and theoretical parameters pertaining to standards and their role in the design process. The second part presents a synthesis of our research, presented so as to reflect upon some of the major issues in the standardization area. Based on our in-depth examination of the relationship between the technical content of voluntary standards and the institutional factors involved in producing them, the chapter suggests several conceptual correctives to analytical frameworks that have been built up around standards over the past 10 to 15 years.

THE TECHNICAL AND INSTITUTIONAL SIGNIFICANCE OF STANDARDS

We must be careful at the outset to clarify exactly what technical standards are, and what they do. Definitions and taxonomies of standards are becoming abundant to the point of confusion (Salter and Hawkins 1990). By the most general and simple definition, however, a technical standard is a point of reference or comparison that has been commonly identified in some way, in order to enable the exact repetition of actions and/or the precise duplication of physical characteristics. Standardization, on the other hand, refers to the process by which standards are established and applied. In practice, however, it is preferable to conceptualize the entire enterprise in terms of an ongoing standardization process, referring broadly to the various circumstances in which standards are proposed, created, used, revised, and abandoned. Indeed, with respect to individual standards and standards frameworks, these dynamics can be conceptualized as a kind of career path that can intersect with various institutional structures at different times.

In broad terms, an institution is understood here to mean a locus of human activity in which patterns of behaviour are established by rules, which, in turn, are legitimated by commonly held precepts and beliefs (Neale 1987). In standards-making, as in most other enterprises, there are both formal institutions, of the order of legally and constitutionally established organizations, and meta-institutions, referring to concepts and belief systems to which members of the formal institutions adhere to greater or lesser degrees—the state, technology, and so forth.

Standards can be established institutionally by *fiat*, as with laws and regulations, or they can result from practice and custom, as with the standard designations for 'left' and 'right'. However, standards can be established also through various forms of interinstitutional consent that may have only oblique relationships to either legal process or cultural tradition. The product of an individual commercial firm, for instance, can dominate a market to such an extent that this product becomes either *the* standard by which all other similar products are judged, or, in extreme cases, a product becomes *the* standard by virtue of the fact that it is the only product of its kind in widespread use. In such cases, a *de facto* standard could be said to exist 'by consent' (even if grudgingly) in that no viable alternatives are pursued.

This chapter is mostly concerned with voluntary standards as established by formal consent. These standards come about through institutionalized practices and procedures, usually involving an inter-mediary negotiating body. The element of consent is normally operationalized in that decisions are made by consensus (although not necessarily by a unanimous decision) of the affected parties.[1] The intermediary bodies can be trade and professional organizations, government departments, and international organizations. More often, however, voluntary standards are negotiated in formal standards development organizations at national and international levels. In most countries, recognized, and often government-supported, standards development organizations assess industry and public requests for standardization, and organize expert committees to draft standards specifications. Throughout the rest of this chapter, the term standards will refer to voluntary standards as developed by consensus negotiation among stakeholders outside of specifically legislative and/or regulatory contexts.[2]

The nature of voluntary standards

On the surface, voluntary standards seem an ideal candidate for the status of public good in that, in principle, their use by one party does not devalue their utility to other parties (Kindleberger 1983; Olson 1971; Samuelson 1954). However, standards also draw attention to a problem with public goods theory as applied to technology, since they seldom exist in a state of independent utility. The ability to deploy them is typically restricted to (often dominant) suppliers. Moreover, standards can be applied to products and services which may not themselves be available as public goods. As Noam points out, the telecommunication system is not a pure public good because 'users can be excluded and charges can be assessed' (Noam 1987: 47).

Furthermore, although in principle the aim of voluntary standardization is to produce standards that will be available on a non-discriminatory basis for 'public' use (Reddy *et al.* 1989), the voluntary-consensus process does not guarantee that the resulting standard will provide benefits in any truly 'public' sense. A voluntary standard may only serve to legitimize the consensus of a limited group, and its application may prejudice technological environments in favour of vested interests (Rosenberg 1976).

The parameters for defining standards and standardization have changed over time to reflect changing institutional relationships. In the mid-nineteenth century, Joseph Whitworth, one of the first prominent engineers to advocate publicly the institutionalized development of voluntary industry standards, defined these standards in terms of:

the attainment of uniformity of size, and interchangeability of parts, coupled with the adoption in every manufacture of the smallest number of patterns and sizes with which the wants of the consumer could be satisfied. (Whitworth 1882: p. iii)

Most definitions proposed by subsequent generations of engineers merely adapted Whitworth's basic precepts to suit the increasingly formal institutional nature of standardization, and the eventual wide adoption of the institutional ethic that voluntary standards are established by otherwise competing industry actors, primarily to yield public benefit in the broadest sense of the term.

In our own day, however, it has become increasingly common for standards to be defined in terms of strategic business objectives. In some definitions, standards appear to have less to do with the rational optimization and exploitation of technology in the public interest, and more to do with co-ordinating technological development with market forces:

Standardization is the product of a personally held belief that the market has the ability to understand and chart a valid future direction through the use of collective wisdom, to understand the impact of change upon itself, and to adjust to that change. The specific change agents utilized in this process are collective descriptions of how things ought to be and function, called standards. (Cargill 1989: 41)

This epistemological and semantic evolution gives some indication of the meta-institutional scope for defining the standards process—the production system, the market, the user, the human dimension, and so forth—and it clearly implies the necessity for formal institutional structures. But what of the functional scope? Functional taxonomies can be expressed in terms of specific objectives—to ensure quality, to define processes, to promote compatibility, to regulate behaviour, and so forth. All standards, however, can be classified more broadly in terms of their fundamental, descriptive, or prescriptive natures (Hemenway 1980; Salter 1988). *Fundamental* standards provide the objective means by which physical characteristics can be measured and assessed. *Descriptive* standards specify particular physical conditions and processes. *Performance* standards specify desired conditions but not the means to achieve these conditions. In practice, all types of standards have institutional affinities that are not strictly technical. MacKenzie, for example, has shown that in some circumstances even the form of basic arithmetic is open to negotiation among various institutional interests (MacKenzie 1993). O'Connell has shown similar phenomena for such fundamental standards as systems of metrology (O'Connell 1993).

In communication networks, all three of the above types of standards apply, although most are descriptive in that they document the technical specifications of physical characteristics and/or operational functions as 'means to ends'. As far as the actual structure and configuration of the network is concerned, there are basically two areas in which standards can be applied. The first is at the interfaces between various pieces and types of equipment. Some interfaces are physical (plug connections), whereas others are logical, defined largely by the operating and applications software. Interfaces, however, only ensure that channels for communication exist. For exchanges on the network to be meaningful, conventions are necessary to format and control the input and output of information such that it is recognizable between the transmitter and the receiver. These conventions are referred to as protocols, and they apply to the syntax and semantics of message elements.

Interfaces and protocols can be proprietary or otherwise limited to specific applications amongst a limited group of users. Essentially, the immediate effect of employing standardized interfaces and protocols that are not proprietary to any particular equipment or service provider is that, in principle, the network is opened up to a wider spectrum of users. Additionally, it is made more transparent for purposes of optimizing existing paradigms of network use, and for designing technical improvements and new use paradigms. For new technology to be usable in an existing network, the design of its interfaces must encompass the existing design parameters of the whole network. Protocols can be more specialized—much in the same way that the same telephone circuit is functional for communication in English, French, Japanese, or any other language recognized by the communicating parties.

Standards and the design of communication technologies

Generally speaking, sociologists of science and technology maintain that the technical and social aspects of design are in a continuous reflexive relationship (Callon 1987). Moreover, speaking as a participant observer, Bucciarelli stresses that every aspect of a designer's environment must be considered for analytical purposes as having the potential to influence the design project materially (Bucciarelli 1984). Even from the perspective of the engineer, Glegg maintains that the designer of technology must operate in three intellectual capacities—as inventor, artist, and analyst respectively—as effective design must incorporate subjective as well as objective criteria (Glegg 1969: 18–19).

At one time, most engineers probably would have placed standardization activities in Glegg's analyst category. Traditionally, in most industries, standards have been regarded as phenomena at the trailing edge of technology—as the consolidation of scientific understanding and implementation experience rather than as the focus of novelty. Nevertheless, existing standards enter into the design process in that often they constitute a major part of the technological foundation upon which new technologies are built, that is, they act as a kind of technical infrastructure in their own right (Tassey 1991).

However, the increasing technical interconnectedness of industrial sectors is yielding a situation in which standardization activity across the industrial spectrum is attached more and more frequently to the R&D process. Thus, standards can also have an upstream function with respect to design, because it is possible to develop new standards in tandem with the new technologies to which they refer. Indeed, in some cases, new technologies are developed with the strategic objective of becoming standards. The case of CD-i (discussed above) is but one of the many obvious examples of this practice that abound in the consumer electronics area. The CD-i situation, however, illustrates *de facto* standardization. Maintaining distinctions between 'standards' and 'products' as separate domains is often difficult with formally negotiated standards as well. Examples are the Integrated Services Digital Network (ISDN) and the Global System for MobileCommunications (GSM). At one level, ISDN and GSM are framework standards, invisible to the user, that have been developed to support digital telephony applications. At another level, they are 'brand names' that identify and promote specific approaches to the provision of services over others.

Standards are a highly problematical design element, however. They are typically not visible to the end-user of a technology except at fairly banal levels. The travelling business person, for example, becomes aware of standards, specifically the problems of incompatible standards,

whenever an electrical plug adapter is required to connect a laptop computer to the mains power supply in a foreign country. This is an inconvenience not without significance in its own right as an indicator of a particular set of entrenched institutional relationships. What our laptop user typically does not see, however, is the multitude of electrical standards on either side of the plug interface, or the many standardized interfaces and protocols in the telecommunication network to which the laptop's communications modem might be connected, or the standards (probably proprietary in this case) that allow the remote user access to the private network services in his or her home office environment.

All of these standards reflect various forms of consent and/or coercion that are worked out between a plethora of institutional interests and there are substantial degrees of freedom in the working out process. As a result, they are significant determinants of what the network technology user can do under what conditions and at what cost. In communication networks, standards embody administrative and economic design criteria as well as technical criteria, and they can embed institutional relationships and patterns of control into the design of the network (Mansell and Hawkins 1992). Furthermore, once standards are implemented, and particularly when applied at critical network inter-faces, their effects are typically spread throughout the whole network, thus often setting up patterns of network development that future innovations must accommodate.

In principle, of course, standards can also disrupt established relationships. Indeed, from both the perspective of corporate network strategists and public policy-makers, this is now often the primary rationale for encouraging the development and use of voluntary standards (see below). Nevertheless, although many institutional entities may feel the impact of standards, there are notable discrepancies and difficulties in the capabilities of individual institutional groupings to influence the content and application criteria of those standards.

One of the problems is that, historically, voluntary standardization has been a supplier-led activity (Hemenway 1975; Sinclair 1969; Thompson 1954). This was partly because suppliers deployed the standards in the first instance, and partly because suppliers had the most to lose if standards were implemented that were inimical to their existing or planned product lines. However, we must recognize also that suppliers are usually in the best position to exploit an inherent division in the standardization process. Although standards mediate many institutional interests that are not technical, they do this within a predominately technical discourse (Majone 1984; Salter 1985; Schmidt and Werle 1992). Indeed, it has been noted that establishing the competency and authority to define standards has been a factor in the historical development of engineering and design as professionally

disciplined activities (Mai 1988). Thus, traditionally, the process itself has existed within boundaries that effectively excluded communities unable to muster the resources to promote their interests effectively in both technical and non-technical discursive modes.

Governments have played a special role in bridging this discursive divide. Individual government departments frequently maintain (sometimes substantial) technical expertise in order to fulfil their general mandates to promote what they conceive to be the public interest in their specific areas of responsibility. Thus, standards have been used in a number of ways to infuse public policy goals into the design of technical systems. Governments can support certain standards or approaches to standards over others. This can be done through participation in the standards process, and through the provision of funds for selected standards initiatives. In general, however, public sector support for standards is normally channelled through procurement policies and decisions (Geroski 1990; O'Connor 1984).

Often underlying these instances of direct political intervention, however, are indirect policy objectives in which the standardization process is expected to act as a surrogate for direct regulation, or as an instrument of industrial policy in its own right (Baggott 1986; Barry 1990; Breyer 1982). In each case, the impact on the design process can be considerable, particularly where it concerns public sector preferences in large national or regional markets for local over imported technologies.

Defining an analytical perspective on standards and design

The complexities of standardization as an element of network design call into question the wisdom of seeking explanations that are rooted in a polarization between instrumental and institutional perspectives on technology. This polarization is evident within the social sciences themselves in the form of the debate about whether technology has an exogenous or endogenous relationship to social change. More significantly in the present context, however, it is this polarity that also largely distinguishes the discourse of the technologist from that of the social scientist.

A purely instrumental analytical frame focuses on determining degrees of utility for actions rather than on the nature of these actions and their antecedents. Applied to standardization, the instrumentalist perspective assumes basically static organizational and economic conditions. Its premiss is basically that under similar sets of conditions, it would always be rational to recognize standardization as an obvious solution to a technical problem. Verman, for example, despite viewing standards-making as a hybrid between engineering, economics, and diplomacy, nevertheless presumes that the success of standards-making initiatives depends upon the closeness of the match between (as he sees it) the rational nature of technical solutions to problems, and the degree of rationality in the organizational and procedural structures employed in standards-making (Verman 1973).[3]

This kind of analysis is obviously very limited in what it can tell us about the underlying dynamics of change. Typically, for example, it might stress the role of standards in co-ordinating technologies to achieve a better 'fit' with market conditions, but not consider the significance of the social and political factors that may have affected the development of the technology being standardized, and the structure of the market in which it will be applied. Moreover, although orientated to the consequences of action rather than to the causes, instrumental analysis also tends to overlook the consequences of any systemic inequities and biases that might exist in standardization mechanisms and processes.

Despite their deficiencies, instrumentalist presumptions not only dominate the philosophies of standards development organizations, but they also permeate virtually every public policy agenda with respect to standards. Furthermore, as the instrumentalist perspective lends itself more easily to formal theoretical modelling, it pervades much neo-classical economic theory. While not ignoring institutional factors altogether, economic modes of analysis of this kind have a strong tendency to conceptualize standards as instruments for achieving optimal technical outcomes in the exploitation of technology—achieving the 'right' standard at the 'right' time according to the prevailing market conditions.

Nevertheless, as part of a broader perspective, an instrumental focus can be valuable in that it offers a constant reminder that whatever other factors standards may or may not embody, at some level they are supposed to function in a utilitarian way as technical devices. Moreover, it reminds us that this is primarily what most technology producers and users expect from standards.

A major analytical pitfall is to assume that because a standard exists, a standardized environment exists also. In the information and communication technology sector, there are often many more agreed standards than applied standards. Frequently, for a variety of reasons, standards are developed that do not actually work, or are never implemented.[4] There are also instances where standardization seems the most reasonable course of action, and yet standards do not appear. Long-standing and notorious examples include the persistence of idiosyncratic national numbering systems, and dial tones (for 'connected', 'engaged', and so forth). It is, therefore, extremely important to be able to evaluate a standardization initiative in terms of the probable utility of its outcome, and never to underestimate the kinds of technical, organizational, and sometimes scientific factors that may not be under the direct control of individual groups of standards developers.

An institutional framework applies dynamic conceptual models of the relationship between technology and socio-economic structures and processes of various kinds. From this perspective, the function of standardization is something to be discovered anew with each significant change in the way these structures interact in the design of technology. The challenge for institutionalists is to reconcile the theoretical position that technology is responsive to socio-economic forces—indeed, that technology is a product of them—with the inescapable observation that although these forces can be highly diverse, variable, and driven by all manner of temporal concerns, the possibilities for technological innovation are, in practice, usually neither immediate nor infinite.

Nevertheless, to assert that technology is the product of socio-economic or institutional forces need not necessarily imply that it is immediately or infinitely responsive, or that the co-ordination problems involved are insignificant. Economists in the institutional tradition, for example, have shown that although technical change is indeed spurred by a variety of social factors, the investment, R&D, and commercial trajectories take time to establish, and, once established, become institutionalized in their own right (Freeman 1994*b*). Furthermore, they have shown that technology developers establish long-term commitments to particular courses of action. These can be expressed theoretically as technology paradigms, technology trajectories, dominant designs, technology regimes, and so forth (Dosi 1982; Nelson 1988; Sahal 1985; Saviotti and Metcalfe 1984), concepts that apply also to many aspects of standardization.

Much contemporary economic analysis of standards reflects this line of thought to some degree. At one time, most economists, like most engineers and industrial project managers, perceived standardization to be an outcome of technology maturation. Standards were viewed primarily as instruments for reducing the variety of industrial components and processes, and, hence, for increasing efficiency, decreasing costs, and enabling the transportability of industrial production capacity from one locale to another (see Vernon 1966). The current view among many economists, however, is that standards are also a factor in the innovation process, and that standards decisions create path dependencies for future technological developments (Arthur 1989; David 1987; David and Steinmueller 1990). To a considerable extent, this evolution in thinking was a product of economic studies of information and communication technology network phenomena in which early or even a priori standardization was often a prerequisite for the introduction of new technology (see Besen and Saloner 1988; Katz and Shapiro 1986).

A lacuna in the economic analysis of information and communication technology standards has resulted from a heavy emphasis upon examining situations of de facto standardization. As David and Greenstein explain, this is something of an inevitable consequence of the difficulties encountered by mainstream neoclassical theory when confronted with the organizational, political, and human dynamics encountered in committee negotiating structures. In these environments, rational choice criteria become harder to identify and control in formal models (David and Greenstein 1990). Some attempts have been made to come to grips with this problem by employing game theory approaches which reintroduce rationality as an abstract element in discrete decision-making situations (Besen 1990; Farrell and Saloner 1988; Swann 1994).

If we look at the history of standards development, however, we see that standardization as an institutional process has been in a state of continual evolution (Weidlein and Reck 1956). For the most part, it has reflected changes in relationships between suppliers and users, and between suppliers of components and integrators of whole products (Hemenway 1975; Sinclair 1969; Thompson 1954). The organizational structures and procedures for standardization are clearly temporal products of institutional alignments as they exist at given points in time, rather than expressions of some universal principle of rationality in assessing and addressing technical problems.

Reddy is correct to stress that voluntary standardization almost always involves informal collaborative action among otherwise competing entities (Reddy 1987, 1990). Reddy conceptualizes the rationality of standards-making in terms of situations in which competitors might see more intrinsic logic in developing technology collectively than competitively.

Standardization . . . is not one specific task, but rather a continuum of collectivity to first define the collective scope of the technology, and second, to reach agreement on the nature and level of interfirm interdependence. (Reddy et al. 1989: 18)

Indeed, it is possible to conceptualize the standardization process in terms of technical diplomacy (Hawkins 1995a). Participants in diplomatic negotiations strive, in the first instance, to determine the exact levels at which agreements to act collectively are possible or impossible (Zartman 1977), and there are signs that this paradigm resonates with the experience of standards professionals. Cargill, for example, maintains that it is not necessarily the main function of a standards committee to produce a standards document at all, but rather to provide a forum for continuous and relatively informal interfirm and interindustry discussion concerning the implementation criteria for new technology (Cargill 1989: 113).

It is important to extend the analytical framework beyond the dynamics of firms, products, and markets. Standards do not simply solve technical and commercial problems. They reflect specific epistemologies, and their discursive elements play a role in establishing the legitimacy of political as well as economic positions by forging links to the presumed objectivity and rationality of technical and scientific discourses (Majone 1984; Salter 1985). In this capacity, standards become important factors in the formation of perceptions as to how technology functions—in a particular instance, or in a society as a whole. Indeed, it is principally by this route that standardization issues become attached to specific commercial and political agenda (Jasanoff 1986; Salter 1988).

A reconciliation of the instrumental and institutional natures of standards is clearly not possible unless we expand our range of vision to encompass the multitude of power structures, and the attendant discourses, that come into play in the standardization process. Actions to co-ordinate technological innovation with various social requirements always involve trade-offs and costs. A comprehensive analysis of standardization must balance the discourses that surround efficiency and utility with those that surround responsiveness and equity.

INSTITUTIONAL BIAS AND STANDARDS-MAKING

Although the role of standards in networks is influenced by the different institutional alignments that may come into existence at any given point in time, not all institutional actors have the same capacity to inject their particular biases into the process. In this second part of the discussion, these interinstitutional dynamics will be illustrated with reference to a number of concrete situations that have been examined in our research on the network environment, the standardization system, and the human dynamics of standards-making.

Standards in contemporary telecommunication networks

It is now redundant to think of the telecommunication infrastructure solely as a conduit for the flow of information. The contemporary infrastructure is made up of a universe of network-based services, controlled for the most part by software applications (Mansell 1990), and intricately related to a 'substructure' of network switching, routing, management, and transmission technologies. Furthermore, future paradigms for network development are dependent on the conditions under which network 'intelligence' is deployed, that is, where in the network the software applications are situated, and who has control over their development, implementation, and use (Mansell 1993*a*).

In terms of enabling networks to operate, our research has shown that the role of standards has changed from the relatively straightforward task of maintaining coherent switching and routing hierarchies at the substructure level to the highly complex role of defining the conditions under which the 'intelligence' allowing for greater flexibility in designing and using network-based services will be deployed and to what effect. The nature of this change, and the institutional realignments it involves, is illustrated by the gradual breakdown of a pair of long-standing commercial and administrative partitions in the telecommunication network.

The first partition is between national and international networks. Historically, in national networks, there were two primary design criteria for the technical interfaces: (i) each new network technology had to be functional when connected to the base of existing technology; and (ii) technical upgrades in a national network had to be consistent with the arrangements by which that network was connected to other networks. Normally, the former requirement was met by the national public network operator (or operators), through vertically structured R&D and commercial arrangements with selected equipment suppliers. It was often debatable whether these activities actually constituted standardization as such, or simply the internal specification of procurement requirements by the network operator.

In any event, such practices meant that typically there were considerable technological idiosyncrasies from one network to the next. To enable communication between networks, there was a need for the formal negotiation of technical and administrative principles to enable national networks to interoperate. This activity became the almost exclusive preserve of the International Telecommunication Union (ITU), and the ITU recommendations became, in effect, the standards for telecommunication technology world-wide. Typically, however, these recommendations would be implemented in national versions, adapted to the prevailing conditions in individual national networks.

The digital network technology that began to make an appearance in the 1970s was far more modular than earlier-generation analogue technology, and, depending upon the functionality provided, individual modules could come from a variety of suppliers, both inside and outside of the traditional telecommunication sector—from computer manufacturers, for example. Furthermore, the amount of data communication carried over the public telecommunication network increased dramatically, thus bringing large computer firms into many of the same commercial and political orbits as the public telecommunication operators. To accommodate these new commercial and service parameters, pressures increased for international frameworks of voluntary standards to intrude deeper into domestic networks.

Digitization and the subsequent entry of new actors and service concepts into the telecommunication sector focused attention on the simple fact that whoever controls the standards also controls the design and use parameters for the network (Mansell 1993a: 15–41). The main issues are: (i) the extent to which it is technically possible to decentralize the intelligence necessary to operate the network; and (ii) the extent to which centralization or decentralization biases the respective abilities of the large public network operators and the suppliers of computer communication facilities to exploit the commercial possibilities of the emerging network environment. To a large extent, both of these issues are contingent upon how the standards questions are resolved and in which sector.

The other longstanding partition is between public and private networks. Specialized private network functionalities are defined either by exchange equipment on the customer's premises, or by specialized technical configurations in the public network.[5] Although it was necessary for the supplier of customer premises equipment to use public network standards for the interface between the private exchange and the public network, the internal specifications of the exchange normally remained under the proprietary control of the supplier. Thus, the user became dependent upon the supplier for upgrades and major recon-figurations of the customer premises equipment.

Our study of the development of standards for mobile telephony, however, yielded strong indications that the basis for maintaining these 'lock-in' arrangements is quickly disappearing. As conceptualized by many network operators, most equipment suppliers, and increasing numbers of legislative and regulatory bodies, mobile telephony is rapidly evolving from 'radio telephone' services for specialist users to mass market applications. However, it was found that in network design terms, the term 'mobile' no longer referred primarily to the type of terminal equipment employed, that is, wireless terminals, but to the physical movement of individual users. Concepts like Universal Personal Telecommunications (UPT) pointed to an integrated design paradigm for fixed and wireless technologies that would allow users to transport their individual service profiles to any piece of available terminal equipment, whether fixed or wireless, and irrespective of national boundaries.[6]

Indeed, the research showed that telecommunication network planners were beginning to incorporate a 'mobility imperative' into the design of the basic user interface with the network (Hawkins 1993a). If universally applied, this imperative essentially would change the overall network paradigm from one in which the user has to 'find the network', that is, physically to locate a network node in order to gain access to the system, to one in which the onus is on the network and service providers to 'find the user' wherever he or she may be located. Such a facility would be dependent upon very substantially increased amounts of network intelligence.

Particularly in the case of wireless technology, it can be seen immediately that the mobility imperative not only widens the scope for the harmonization of standards in both public and private networks, but also threatens traditional regimes of institutional control over these networks. Modern digital wireless telephony distributes some of the 'intelligence' in the system to the terminals—operations like user identification and service level authorization are carried out through an exchange of information between the terminals and the fixed parts of the network. The very physical portability of the wireless terminal makes it unlikely that users would see any advantage in acquiring two pieces of otherwise identical technology, one configured for the public network and the other configured for the private network, or, for that matter, separate pieces of equipment for domestic and international use.

By examining the circumstances that lead to the current blurring of distinctions between the national and international, and the public and private network environments, our research further demonstrated the redundancy of the term 'conduit' in describing network functionalities. The amount of service flexibility allowed by the network (in terms of mobility, for example) is clearly dependent on the amount and nature of the 'intelligence' available in the fixed network. The transport-ability of service functions across national boundaries is dependent either on the availability of similar intelligence capabilities in these various national domains, or on the capability to bypass under-developed national networks entirely. This intelligence, however, can be distributed between public and private network domains. Thus, in a network involving substantial integration of network elements, the pressures from users and new market entrants for a transparent and harmonized structure of global standards are much greater than they would be were these environments to remain structurally separated.

The example of standards for mobile telephony illustrates an essential interinstitutional tension in the information and communication tech-nology design process especially well. For equipment suppliers, particularly those outside the traditional telecommunication sector, there is an undeniable technical and service 'logic' to designing according to common standards. On the other hand, established monopolistic public network operators often see more 'logic' in using restrictive standards regimes in order to continue to control the ways in which the network is developed and exploited, thus dictating the terms on which new entrants can gain access to the telecommunication market-place. This kind of tension presents major challenges for the international system in which standards are developed.

Standards and the mediation of institutional biases

The observation that the historical partition between national and international network standards appeared to be disintegrating, highlighted a phenomenon that became a focus for our research into standards. This was the seeming migration of the locus of institutional power in standards-making from national public network operators and the ITU, to regional standards bodies, surrounded by a plethora of formal and informal industry consortia of various descriptions (O'Connor 1992). This highly visible change in the institutional landscape proved to be the 'signal' indicator of a vast array of sometimes less visible, but often more significant changes.

The regional emphasis began in the early 1980s. A primary stimulus was undoubtedly the requirement to maintain the technical integrity of the public network in the three largest telecommunication markets—the USA, Europe, and Japan—where the established monopoly structures were beginning to be challenged. In these circumstances, the ITU mechanism for issuing recommendations was considered to be too politically cumbersome, too isolated from the pace of technical change, and too remote from the immediate needs of network designers in these three individually large markets. Thus, in 1983 shortly before the divestiture of AT&T the US T1 committee was formed.[7] In 1985, spurred by increasing liberalization initiatives in Japan, the Japanese Telecommunications Technology Committee (TTC) was founded. Finally, in 1988, the European Telecommunications Standards Institute (ETSI) was established, largely in response to a series of initiatives by the European Union to encourage the development of a pan-European network environment, based upon harmonized standards (European Commission 1987).

The primary significance of regional standards bodies for the design of networks, however, was found in our research not to be technical or geographical but political and procedural. This new regional standards focus must be seen against the backdrop of changes in the traditionally protected relationships between operators and suppliers, and the entry of new actors into the telecommunication sector (Hawkins 1992, 1993*b*).

Direct participation in telecommunication standards-making had been previously closed to all but the national public network operators, albeit often with input from preferred (usually domestic) equipment suppliers. However, the costs of engineering and deploying digital telecommunication technologies were very high, the range of required technical capabilities was burgeoning, and the life-cycle of the new technologies was comparatively short. These factors, coupled with the prospect for manufacturing firms and service providers alike that new footholds might be obtained in previously restricted international markets, made it virtually certain that, at some point, the standardization process would be opened up to participation by all materially interested parties. The regional bodies adopted participatory criteria and administrative procedures that brought standards-making for telecommunication into line with practices for setting voluntary standards in most other industries.

All of these manœuvres were undertaken with the stated objective of ensuring that standards development would become 'market-led', but it is not certain that this is what has occurred. Our studies of mobile communication systems, for example, indicated just how strongly standards become linked with particular product and service development strategies at design stages that are far upstream from the process of identifying specific market requirements (Hawkins 1993*a*). Moreover, trade and industrial policy considerations could be identified as driving forces in the development of standards in regional institutions (Hawkins 1994*a*). Furthermore, our analysis of user influence in standardization revealed that there are few effective mechanisms, regional or otherwise, by which users can inject their own requirements assessments into standards projects at upstream or downstream stages (Hawkins 1995*b*).

Standards and the relationship between producers and users

One of the central precepts explored in this book is that the relationship between producers and users of technology in the design process cannot be explained by the rhetorics of utopia and dystopia—that the interaction can be subtle, and its positive and negative implications counterintuitive. On the surface, the standardization process seems to offer an ideal setting for users and producers to co-ordinate their perspectives and agenda with respect to technology. The reality is more complex.

The conditions that resulted in the restructuring of telecommunication standards organizations, also placed pressure upon this new system to permit direct participation by all interested parties, including user firms and organizations. It has proven very difficult, however, to incorporate users into the process. Their participation rate in Europe, for example, typically stands at less than 2 per cent of the total number of person/ committee hours.[8] Our research showed that users were only beginning to see that the effective use of communication technologies might at some point involve them in design decisions, including standards selection. The distribution of standards initiatives among a proliferation of organizations, however, made it less likely that user firms could justify expending the resources necessary to ensure user-friendly outcomes.

Some institutional economists have explored situations in which the design capabilities for a technology are distributed to varying extents between producers and users, with products resulting from interactive design processes involving both entities (Lundvall 1988; von Hippel 1977). Others, however, have noted that this model is troublesome with respect to standards as the initial range of technical choices available to the user tends to be already circumscribed by technology producers (Dankbaar and van Tulder 1991). Our own studies indicated that the benefits envisaged by producers from greater user involvement in standards for communication networks were very much more one-sided than the kind of creative user–producer symbiosis that is often expounded in the design and innovation literature (Hawkins 1995b). Suppliers of network equipment and services tend to view standards as ways to develop markets for broad new technological environments whereas users are more typically concerned to have standards that will rationalize and maximize their use of existing environments, or support specific new functionalities. Producers are still prone to let technology lead the service environment whereas users want functional requirements to lead.

Although both users and producers have a stake in standards-making in the communication sector, the relationship between them is not essentially symbiotic. Both producer and user interests in standards are concerned with gaining advantage over each other and not with some abstract notion of increasing technical efficiency. Moreover, it is evident that the user profile with respect to standards is evolving quickly. As the system of networking becomes more decentralized, users have begun to acquire a greater degree of parity with the suppliers in their ability to specify and configure many of their own requirements internally. The voluntary standards mechanism, however, remains orientated primarily to addressing the traditional requirements of dominant producers. The sustainability of the current structure for information and communication technology standards development organization will depend in no small measure upon its capacity to respond to this tension.

Standards and the politics of network design

The problems inherent in using voluntary standards for regulatory and policy purposes have been mentioned above. Many of these were examined in our work by focusing on the evolution of standards policy in the European Union. We examined the interrelationship between European Union standards policies directed at the unification and liberalization of European telecommunication equipment and service markets, and policies supporting the convergence of the telecommunication and information technology sectors.

The concept of a pan-European network environment as envisaged by European Union policy-makers has presented a problem of political as well as technical design. A large part of the European Union strategy in this area was to use standardization as a mechanism to establish distinctly 'European' network design trajectories. Thus, many of the resulting standards became technological expressions of political ideals. Nowhere was this more evident than in the field of mobile telephony. A mobile terminal that could be used in all of the European Union member States, thereby circumventing to some extent the restrictive administrative regimes of the national public networks, was an attractive political symbol of the kind of free movement of people, goods, and services envisaged by the architects of the 'single Europe'.

Attempts to implement such systems, however, illustrate clearly the distance that can exist between political aspirations and the realities of established industrial structures. Only with GSM did policy support appear to play a relatively unambiguous role in the widespread acceptance and deployment of a new network environment based upon a distinctly European standard. In this case, however, consensus on the need for a pan-European system had been forthcoming from the national monopolistic public network operators well in advance of the period during which GSM became attached to an overt political agenda in the European Union to liberalize telecommunication markets. In subsequent cases, public and private sector agenda have been less well co-ordinated, and the use of standards in the absence of complementary policy instruments of a more formal kind, has not overcome political, industrial, and administrative rivalries between member States.

There can be serious consequences if standards policies are not based upon an assessment of the particular complexities of the standardization process. European Union standards policy, for example, has consistently followed a strictly instrumentalist line. The assumption has been that the process is essentially linear and that its inefficiencies could be addressed with organizational and procedural changes and with 'pump priming' in the form of European Union funding for selected standards programmes, and standards-based procurement policies (European Commission 1990).

Thus, European Union attempts to restructure the voluntary standards process for political purposes were problematical. Typically, they were at odds with what participants expected from the process, and, correspondingly, with the levels of commitment they were prepared to make to developing European standards. Frequently, the result was confusion in Europe, and internationally, over the voluntary or mandatory status of 'European' standards. Moreover, it was found that confusion among standards-makers and standards development organizations as to the political or technical motivations for European standards initiatives often resulted in lack of confidence in technical and commercial circles alike as to the suitability and/or quality of the standards produced (Hawkins 1993*b*).

Standards and the international politics of network design

Given the assertion that communication systems cannot function without standards, it might be presumed that international systems cannot function without truly international standards. The geopolitical locus of power in the global telecommunication system, however, appears likely to remain the nation-State for some time to come (Hawkins 1994*a*). Despite frequent assertions that the communication network is 'globalizing', that is, becoming increasingly delinked from the control of individual nation-States, so-called globalization tendencies in telecommunication have to be viewed against the continuing background that a relatively small number of States comprise not only the bulk of the telecommunication market, but also the bulk of the R&D capability, and the corresponding capacity to influence the selection and application of standards.[9]

Not surprisingly, the restructuring that is occurring in the international technical institutions for telecommunication in the light of these globalization tendencies, has begun to attract the attention of political theorists (Cowhey 1990; Krasner 1991). Cowhey, for example, conceptualizes the telecommunication standard as a technical mechanism for the sustenance of a regime centred in the ITU that functions as a global cartel of public network operators.[10] The 'regime' is a concept in international relations referring to the ways in which nation-States co-ordinate their international activities by means of epistemic communities formed around various 'issue areas' (Kratchowil and Ruggie 1986). Genschel and Werle posit a variant of this analysis; currently observable turmoil in the organization of standardization in the telecommunication industries reflects attempts by the established industry actors to use standards in order to redefine traditional control hierarchies as national monopoly structures give way to liberalized markets (Genschel and Werle 1992).

This kind of analysis has a number of shortcomings, however, most of which stem from a lack of precision in describing industry structure. In examining the transition from seeming conditions of monopoly to seeming conditions of liberalization, for example, these commentators tend either to overlook the condition of oligopoly as a structural feature of the telecommunication industry, or to underplay the antecedents and current significance of oligopoly. Another difficulty with this form of analysis is that it tends to treat the telecommunication sector as an entity in its own right to the detriment of emphasizing the importance of the links that exist with other industrial and political agenda. A case in point is the perception of the consequences of market liberalization in telecommunication for the traditional relationship between the monopoly national public network operators and the ITU. The corollary would seem to be that a breakdown in this relationship should result in a more open and equitable international system of technical and administrative regulation in the network. Our research confirmed that the ITU is indeed scrambling to redefine itself in a changing regulatory and commercial environment. However, no indication of any kind was found that a weakening of the position of the ITU as a focus for the monopolistic aspirations of national operators had resulted in any corresponding weakening of the oligopolistic nature of the telecommunication industry with respect to the ways in which standards are used to establish commercial and technical design trajectories upon the public network.

Decentralization of the technical standardization function may result in more official openness on the part of standards development organizations to the needs of heretofore disenfranchised actors or new market entrants. But openness does not necessarily result in more responsiveness or equity when addressing their requirements. In fact, decentralization creates special problems of its own (Noam 1994b) Currently, new actors are faced with the prospect of monitoring a bewildering array of mostly disconnected standards bodies at national, regional, and international levels. Few except the very largest firms can hope to sustain a level of participation in standards committees that is sufficient either to construct an overall picture of the general directions being taken in any given period, or to intervene in a strategic way. They simply do not have the requisite capabilities.

The regional phenomenon evident at the time Genschel and Werle were writing was also shown only to be a transitional phase in a much more significant realignment of powers. Regionalism as embodied in the early relationships between T1, TTC, ETSI, and the ITU, soon gave way to a sort of *interregionalism* in which the global aspirations of dominant multinational service and equipment suppliers began to be worked out in relation to emerging national postures in international trade (Hawkins 1994a). Beginning in 1989, the regional bodies started a series of interregional summits, ostensibly to co-ordinate regional inputs into the ITU. Participation in the summits was originally restricted to the regional bodies and the ITU. In subsequent summits, however, the participatory scope has been expanded to the extent that future events will be gatherings of select individual national groupings, mostly from countries with already highly developed telecommunication systems, as roughly aligned to the orbits of one or more regional standards development organizations.

The ITU has responded in turn by opening up its membership to individual firms and organizations, resulting in what could be a redefinition of international standards as an institutional construct. The political aspect of telecommunication standards in the ITU, traditionally concentrated on the preservation of national authority to control development and use of network facilities, is becoming redefined in terms of facilitating relationships between the trade strategies of individual States and the productive capacity of multinational firms based in these States.

At the end of the 1980s, Europe was seen as the driving force in regional telecommunication standards. Many viewed the possibility of a European success in achieving a pan-national network environment based on transparent and harmonized standards as placing Europe potentially in the most opportune position to assume control over any future multilateral initiatives that might arise respecting the global development of the telecommunication network. However, US interests soon followed suit with their own initiatives to harmonize the network environment in North, Central, and South America, and the Caribbean, according to US standards. This initiative was launched partly through the auspices of the Organization of American States, a longstanding instrument of US foreign policy in the region (Slater 1967). Likewise, although their plans are neither advanced nor clear, Japanese interests have begun to examine the Asian markets in similar terms.

If we examine carefully the US position in this context, we might consider that we are witnessing less the organization of a new international telecommunication regime, and more the beginnings of a new form of hegemony in which the nature of the bargain between the hegemon and its acolytes is as yet unclear.[11] By supporting a more informal system of technological alliances than that of Europe, the USA has the advantage of being able to act simultaneously as the centre of an emerging telecommunication region, and as an independent State capable of co-ordinating the development of a regionalized technological infrastructure with its own broader foreign and trade policy goals. The European Union has not been able to demonstrate any countervailing ability to co-ordinate its efforts in this way, and the formal nature of the technological and administrative organization of the European telecommunication region may now be the source of its principal disadvantage in world markets.

The practice of standards-making

Basically, there are three categories of institutional participant in any standards initiative: (i) participants already possessing products in the specified area; (ii) participants planning or developing products in this area; and (iii) participants not involved directly in developing products, but nevertheless desiring to exert control over the environment in which products will be deployed. Correspondingly, there are at least three sets of reasons to participate, and three types of expectations as to what the process should yield. In all three cases, negative as well as positive participatory rationales can coexist—some participants may want standards right away, whereas others may want to prevent them altogether, or otherwise to control their content and pace of development.

All the above factors become operationalized in one way or another during the actual process of negotiating standards in committees. It is only at this point that the interactions between the instrumental and institutional elements of standardization can be observed directly. The human perspective is extremely important, however, as it is not institutions, as such, that attend meetings of standards committees, or that develop and apply standards strategies. These functions are performed by standardizers—individuals who represent these institutions, but who may be attuned to specific institutional agenda to greater or lesser degrees at different times. Indeed, it has been suggested that the main problem with the telecommunication standardization process, irrespective of the particular institutional alignments that may apply at any given time, is that the standardizers acquire a measure of autonomous power to direct the process that may or may not be representative of the positions of the institutions they represent (Wallenstein 1990).

Both quantitative and qualitative methodologies have been applied at various times to the analysis of standards committees.[12] In each case, however, the units of analysis have tended to be individual standards initiatives in individual institutional settings. Our work led us to conclude that standards-making for information and communication technologies takes place in a very dynamic milieu that incorporates both technical and non-technical factors in a complex of interrelated projects, institutional structures and organizational contexts. In this essentially multidimensional and multi-institutional environment, individual initiatives acquire significance only in so far as they can be related to a much larger structure. This final section outlines some of the principal characteristics of this *standards milieu* as an institutionalized structure and process in its own right, based on first-hand observations of standards committees.[13]

Among the first phenomenon to come to light was that the dynamics of standards-making actually militate against mobilization of the process towards the attainment of specific objectives according to the agenda of any one individual participant, or subgrouping of participants. Even though, in general, the process was seen to become more biased towards certain institutional groupings than others, it was usually questionable whether any particular actor could be said to be controlling a standards committee at any given moment in terms of specific inputs and outputs.

In the first place, the quantity and diversity of participants have increased with the regionalization of standards activities, resulting in frequent intersectoral rivalries and intensified difficulties in securing consensus. Also, standards initiatives are now frequently pursued simultaneously in several fora. This practice is not always an exercise in duplication—sometimes different organizations have specific competencies in selected aspects of the same problem, and at other times organizations adopt different institutional or technical perspectives. At the micro-level, there are frequently large discrepancies in the attitudes of individual participating entities (particularly firms) towards co-ordination of their standards activities with their overall commercial and political objectives. Some participants co-ordinate their positions strategically in terms of technical and/or non-technical objectives. Others get involved in standards-making on a comparatively *ad hoc* basis. Moreover, the really important exchanges of information often occur through informal, 'off-the-record', and sometimes surreptitious encounters between individual participants.

These dynamics suggest that the outcomes of a voluntary standardization process might be random. However, there are two overarching factors that impose a certain measure of co-ordination upon this process. The first, and most obvious, is that irrespective of how standards activities are managed at the micro-level, that is, in firms and individual organizations, control over the distribution and co-ordination of standardization work within the standards milieu is largely exerted by the standards-makers themselves.

In fact, reflecting the comment above concerning the resources necessary to monitor an increasingly decentralized standards structure, this kind of control masks a structural inequity. The relatively open dynamics at the committee level are by no means an indication that the process was equally responsive to all participants at the institutional level. The networked nature of information and communication technologies means that virtually every standard is related in some way to virtually every other standard. Some standardizers, notably those representing small producer firms and users, are at a major disadvantage in that the diverse and decentralized nature of the standards milieu makes it very difficult for all but the representatives of large equipment and service suppliers to contextualize the work of an individual committee in terms of the status of its subject-matter within the standards milieu as a whole. Indeed, it is most often via key individual professional standardizers, usually from large supplier firms, that day-to-day co-ordination between the many related projects and venues in the milieu is effected.

In some circumstances the institutional nature of the milieu itself can result in incidences of relatively autonomous action by standardizers. Instances were observed where one technical approach was pursued over another simply because of the interests and competencies of members in particular committees, or because of prior commitments by standards development organizations to particular development frameworks. That the process is indisputably difficult to control in terms of pursuing individual agenda, however, does not indicate that the process, in general, occurs without direction from its participant constituencies. By far, the majority of the factors discussed in the standards committees observed in our research, could be externally corroborated at some point with commitments made to individual standards initiatives by firms, industry organizations, or government bodies.

In fact, the second co-ordination factor is the common recognition among standardizers of the need to address real technical issues, i.e. those issues commonly perceived by the standardizers to be at the top of the technical agenda for network design and operation. Most contemporary information and communication technology standardizers (certainly the key people as described above) are engineers, connected to their firms at project management levels related to the design and/or marketing of products and services. Although standards-making is certainly not a purely technical process, our research showed that it is a mistake to underplay the motivational significance of technology as a meta-institutional concept of particular significance to engineers and designers. The technical function of standards affects their selection process, and hence the choice criteria for new design features in the network.

In other words, there is a certain level at which standardizers become designers of network technology, and their primary concern becomes the technical utility of standards in ensuring the interoperation of networks. Indeed, participant observation revealed that there are distinct, albeit non-linear, stages in the standards process, each largely informed by the relative degree of discursive interplay between the technical and non-technical factors.

The early stages of the process are devoted to the discovery of technical issues and problems, and are carried out in a predominantly technical discourse. The middle stages become more concerned with establishing the kind of benchmark that is required in order to determine the logical limits of any further collective action. In this case, the principal concerns were found not to be technical at all: indeed, confusion over non-technical issues was observed to cause more delays in the process than any technical impediment. At this critical middle stage, standards initiatives can fail outright. Alternatively, they can shift their locus from one organization to another, or their subject-matter can become grafted on to the agenda of another committee. The initiative can also enter a kind of 'limbo' in which the subject is kept under discussion but never resolved. Should an initiative become mature enough to result in a document, however, the final stage involves painstaking scrutiny of the technical factors. At this point, the technical discourse dominates. Eventual assessments of the relative benefits and disbenefits of a standard in economic or political terms are dependent on a clear understanding of how the standard is likely to perform once implemented.

CONCLUSION

This discussion suggests that each of the stages in the development of a standard, regardless of the order in which they occur, are very much affected by the degree of co-ordination that various representatives of the interests of institutions supporting particular network design paradigms are willing to accept between a number of often quite independent institutional and technological orbits. The central tension in the organization of standards-making is between the characteristics of the process as it exists at any given time, and the expectations that the participants in standards-making may have of the process at that same given time. As participant expectations reflect different histories, standards-making is an example of the dynamic formation of capabilities as discussed in Chapter 1.

An examination of the changing institutional relationships in standardization yields three main observations. First, although the individual technical aspects of standards are discussed in individual committees, these activities have meaning only when placed in the context of the whole complex of standards-making structures world-wide. The telecommunication sector is still oligopolistic—the majority of the market for equipment and services is still in the hands of a few dominant companies with global ambitions (Bauer 1994; Trebing 1994). These actors currently maintain a presence throughout the standards milieu. To concentrate analytically upon specific cases of standardization, isolated from the whole network environment of services and technical substructure, and/or isolated from the expanding universe of institutions and vested political and economic interests in this sector, is to invite serious misconceptions about the practice and significance of standards-making.

Second, standardization in a dynamic sector like information and communication technology is not simply a matter of making objective choices from among mature or maturing technical alternatives. It is a process through which industry and public sector actors negotiate the distribution of the costs involved in co-ordinating maturing technological profiles with commercial and regulatory strategies. Standards-makers must consider their choices in relation to the already established economic and technical priorities and trajectories of their particular firms and organizations. Thus, standards-making is not so much a process of negotiating changes in the initial positions of participants as it is of discovering where common levels of agreement are possible, given that the scope for flexibility in the technical and/or commercial profile of individual participants with respect to the standard may be minimal.

Third, the complex dynamics that can be observed in the standards milieu highlight the problems inherent in incorporating voluntary standards into the public policy process, particularly where standards are forced into the role of surrogate policy instruments. Standardizers concentrate upon the instrumental side of standards. Policy-makers must balance this with institutionalized perspectives that are themselves subject to change. Policies for voluntary standardization are important support mechanisms in an industrial society, but they cannot be expected to compensate for the inadequacies or absence of more formal modes of economic and industrial policy. Where such expectations are present, the voluntary nature of the standards can be compromised resulting in a lack of industry commitment to them, international suspicion of the motives for standardization in the first place, and the eventual weakening of effective public sector influence over the design of new network environments.

Standardization activity is a mirror of evolving relationships in the telecommunication equipment and service sectors. The activity is primarily directed at determining a common understanding of which aspects of a technology will be exploited in a collective way, and at what point the majority of institutional actors would commonly recognize the redundancy of competitive action. The development of information and communication technology standards illustrates in a particularly cogent way the gulfs that can exist between preferred technological outcomes, that is, outcomes allied to the social, political, and economic agenda of specific institutional groupings, and 'virtual' technical imperatives—in this case, the necessity to maintain a semblance of common order in a technical infrastructure comprised of highly dynamic but interdependent elements.

NOTES

1. The subject of consensus is probably the most misunderstood aspect of standards-making. In practice, it seldom refers to unanimity but rather to the lack of sustained opposition. Where votes are necessary, most standards bodies specify only that a majority substantially greater than '50% plus 1' is required. The International Organization for Standardization, for example, specifies a 71% majority as indicating consensus.
2. In the past, such standards were often referred to as *de jure* simply because they were the result of formalized processes of negotiation. Some writers now limit their use of this term to describe those standards that involve government action in some form, preferring the terms, committee standards, or institutional standards for the products of standards development organization processes.
3. L. C. Verman was at one time President of ISO.
4. A good example of this phenomenon is the Open Systems Interconnection (OSI), a framework of international standards for computer network interconnection. Although dozens of OSI standards exist, only a small fraction of them have been implemented, and estimates of the OSI share of the computer networking market have never exceeded about 6%. Many of the standards are not implementable—either they refer to obsolete technology or they are too 'theoretical' for practical applications.
5. Provision of private network functionality using public network equipment is usually referred to under the general term 'Centrex' Where localized Centrex services are linked together, Virtual Private Network (VPN) services can be provided that include long-distance voice and data transmission.
6. The UPT concept began in the ITU in the mid-1980s. The concept not only involves the integration of all service environments (wired and wireless) through the use of a single customer number, but it also involves charging on the basis of the number rather than on the basis of the particular technical environment(s) providing individual types of services.
7. T1 is an accredited standards committee of the American National Standards Institute (ANSI), the organization that oversees the voluntary standards development system in the USA.
8. This figure was derived from participation data given in ETSI annual reports from 1990 to 1992. The ETSI annual report for 1993 abandoned the practice of reporting these comparisons.
9. According to figures compiled by the Yankee Group, the top forty international telecommunication carriers generated approximately $US 255 bn. in total revenues in 1993. Approximately 65% of these revenues were generated by companies in just five countries (the USA, Canada,

France, Germany, and the UK), see Finnie (1993). By 1990, the total world-wide production of telecommunication equipment amounted to over $US 63 bn. Nearly 75% of this production was centred in seven countries (the USA, Canada, France, Germany, the UK, Sweden, and Japan), see United States International Trade Commission (1991).

10. In his argument Cowhey frequently juxtaposes the terms 'cartel' and 'monopoly'. This is problematic in itself as, virtually by definition, true monopolies do not operate cartels. Strictly speaking, cartels are products of oligopolistic structures in which a market is dominated by a very small number of suppliers who contrive to limit the supply of goods in order to keep prices artificially high.

11. Gilpin (1987) describes the conditions for hegemony as currently understood with reference to the USA. He presents the hegemonic relationship as one in which the hegemon agrees not to seek maximum economic advantage in particular situations in order to acquire control over a more stable structure of external political relationships that may or may not result in material benefit to the hegemon in the longer term. Reference is made to the US position in the world monetary system in the immediate post-World War II era.

12. Salter (1988) employs a qualitative approach based on interviews and analysis of documentation. Weiss and Sirbu (1990) employ a survey questionnaire and statistical analysis techniques. Reddy (1987) employs qualitative participant observation techniques to frame hypotheses that were then tested using statistical techniques.

13. Although these ethnographic studies were conducted and documented rigorously, they were not originally intended as a contribution to a sociology of standards committees. Rather, their role was to provide guidance in posing questions of a political and economic nature for the thematic studies as synthesized above.

7

Network Governance: Designing New Regimes

ROBIN MANSELL

[W]ill the right amount of new information be created, and at the right times? [W]ill the new information that is created be used productively, that is, in a way that yields the maximum flow of social benefits for the producers and consumers of goods and services?

(David 1993: 24)

INTRODUCTION

The design of information and communication networks including both their technical and organizational characteristics is a crucial determinant of whether stocks of codified knowledge can be accessed and used in a variety of electronic forms. It is also an important determinant of whether appropriate incentives are in place for the generation of information and for the socially and economically beneficial use of electronically codified knowledge. We have shown how the capabilities of the information and communication technology and services producers, and business and consumers users, are being transformed through their interaction with innovations in technologies and organizations. The degrees of freedom in the extent to which they, together with policy-makers, are able to manœuvre within the boundaries of technical and structural constraints and to modify their electronic environments vary substantially through time.

The design of information and communication technical network 'substructure' is strongly influenced by the activities of the formal institutions which comprise national and international policy and

Dr Puay Tang, Research Fellow, Science Policy Research Unit, contributed substantially to an earlier draft of this chapter. Professor Liora Salter, Osgoode Hall Law School/Faculty of Environmental Studies, York University and Director of the Canadian Institute for Advanced Research Programme (CIAR) on Law and the Determinants of Social Ordering provided opportunities to present the arguments at seminars and her comments are gratefully acknowledged.

regulatory environments.[1] The interleaving of technical and non-technical socio-economic factors within and among such institutions affects the promulgation of policy frameworks. The evidence in the preceding chapters suggests that negotiations among actors in the public and private sectors may be giving rise to a potential weakening of effective public control over the design and terms and conditions of access and use of new electronic networks and services.

We look more closely in this chapter at these changes in power relationships as they are articulated through the design of new forms of electronic network governance.[2] Governance in this context refers both to the institutionalization of various forms of rule-making and rule-enforcement as well as to the formal institutions which engage in these processes. The international regime which enables public and private sector actors from different countries and regions to construct the communication network substructure on a global basis and to provide services, seeks, on the one hand, to achieve co-ordination on the part of network designers and users; and, on the other, to strengthen the competitiveness of companies located within their boundaries. The contradictions between the need for co-ordination and for competitiveness, and its expression in the relative openness or closure of the network substructure and services, are especially visible in two key areas of the information and communication network governance regime. The first is the design of policy institutions that are concerned primarily with the development of network substructure architecture and operation. The second is the design of institutions that are concerned with the protection of intellectual property and with securing adequate returns for the originators of the electronic information that is conveyed by global network substructures.

A new international regime of network governance is emerging. The formal institutions which have provided the fora for the negotiation of norms, rules, and codes of conduct in this area are finding it necessary to reconfigure their goals and modes of operation and new formal institutions are being created. The result is a new and complex terrain upon which the technical and non-technical aspects of network and service innovation are being negotiated. The early 1990s have seen the partial convergence of the international governance of global information and communication networks in a new formal institution—the World Trade Organization (WTO). The past decade also has seen growing challenges to the International Telecommunication Union (ITU), which historically has provided the governance regime for international telecommunication, and to the World Intellectual Property Organization (WIPO), to find effective ways of resolving tensions between the goals of co-ordination and competitiveness in the development of the network substructure and services.

These developments raise important questions. Are these developments likely to achieve the goals of leveraging open markets and stimulating competitiveness at the same time that they create the conditions for improved co-ordination and for greater accessibility and use of networks and services? As the World Telecommunication Advisory Council to the ITU has observed: 'telecommunications is a fundamental infrastructure—perhaps *the* most important infrastructure of the information age. As such it is an essential public concern and cannot be divorced from basic issues of social equity, or from government regulation' (International Telecommunication Union 1993: 14). The international governance system for communication networks and services has an important bearing on the quality, quantity, and accessibility of the electronic information that is produced and used around the world. It affects how and by whom such information can be used.

The careers or life-cycles of the formal institutions of governance of information and communication technologies are subject to considerable flexibility and to the path dependencies resulting from their respective histories. These institutions do not experience a straightforward linear process of birth, maintenance, and death. Their longevity, functions, powers, and perspectives are influenced by their capacity to address and resolve issues relating to changing perceptions of the appropriate development of markets for both network substructures and services. As such, they are concerned with issues of technical innovation, and economic and social equity problems. This chapter focuses on the uncertainties and ambiguities that characterize the future prospects of the ITU, the WIPO, and the newly emergent WTO. The ITU and the WIPO have relatively long histories. The WTO has only barely begun its career as a formal institution.[3] However, its establishment with the culmination of negotiations within the framework of the General Agreement on Tariffs and Trade (GATT), provides us with an opportunity to consider its prehistory. We look particularly closely at the degrees of freedom available to the designers of an international governance regime that will need to address these co-ordination and competitiveness issues well into the twenty-first century.

INNOVATIONS AND INSTITUTIONAL INSTABILITIES

Fears about the consequences of the declining competitiveness of the suppliers of new generations of information and communication hardware and software systems and services in the major trading regions of the world are fuelling debates about the need for change in the international governance regime. The fusion of network infrastructures with computing and audio-visual technologies has brought a vast range of new information services into the limelight. Multimedia services, especially, are expected to represent a potentially vast market providing information for business presentations, training, and leisure entertainment via desktop computer data and voice communication services as well as via television-based interactive services. The private sector players are positioning themselves to claim a share of potentially lucrative local and global markets for the electronic information services that will rely upon an ever-expanding network infrastructure.

The idea that economic trading regions can fall behind in their capacity to produce and use communication and information services is becoming a familiar theme in the rhetoric of public policy and in the strategic pronouncements of companies based in the USA, Europe, Japan, and other Asia Pacific countries. Voice, text, data, and video services have come to be regarded as among the leading contributors to economic growth and, in many instances, they are expected to generate new employment opportunities (see Freeman and Soete 1994). They have much potential to stimulate both the creation and use of knowledge that is expected to enhance the productive capabilities of economies in every country and region of the world.[4] In consequence, no major trading region wants to see its capabilities in the information and communication field eroded or left undeveloped. In Europe, for example, the notion that 'the European operators will distribute multimedia applications developed by US and Japanese companies on US equipment and with US software' (Commission of the European Communities 1992), is one that resonates with the positions of policy-makers in other regions and which guides their interests in the design of the evolving network governance regime. In this context, the focus over the past decade has been mainly on the prospects for 'making a business of information' (Cabinet Office 1983).

Intermingled with the high profile given to issues of competitiveness and the employment-generating aspects of participation in the production and consumption of information and communication technologies and services is an equally longstanding concern about the social and cultural implications of 'wired cities' and the electronic global villages that are envisaged as a result of the widespread diffusion of advanced information and communication services (Dutton *et al.* 1987; Nora and Minc 1980; Porat and Rubin 1977). The promise of these technologies in the 1990s is one of equitable electronic interactive participation in businesses and public life, of the electronic delivery of public information, and of new forms of teledemocracy (Dutton *et al.* 1994).

The historical diffusion and use of electronic information services and their substructural support networks has been uneven and the prevailing network governance regime has been unsuccessful in creating incentives that would alleviate the gaps in the accessibility of advanced networks and services and the exclusionary consequences that these have engendered. Is there any reason to expect that the instability that presently characterizes the network governance regime will create conditions that will enable more inclusive participation in the production and consumption of electronic information services in the twenty-first century?

Research on the international governance regimes that affect the development of information and communication services has been characterized by a tendency to look separately at the substructure of information and communication networks and at the information content. Cowhey, for example, has examined the telecommunication regulatory substructure regime from the point of view of those who have been winners and losers in the face of the introduction of competition into national markets (Cowhey 1990, 1993). His work has suggested that large corporate suppliers and users of telecommunication services have been the winners. The changes in policy and regulatory regimes have not significantly enhanced prospects for the alleviation of the exclusionary biases in network development. He also has argued that the negotiations on trade in services which took place within the GATT framework between 1986 and 1993 provided a forum for controlled conflict between the older ITU and newer GATT governance regimes (Cowhey 1990: 197).

Drake and Nicolaidis (1992), in contrast, have turned their attention mainly to the governance regime for information services or content. They regard the rise of a new governance regime for information and communication services as the result of changes in the broader context of the globalization of financial and other service markets (Drake and Nicolaidis 1992). The emergence of *compaks* (complex packages in which goods coexist with services) is blurring distinctions between the production and consumption of information (Bressand *et al.* 1989), as we saw in Chapter 5. The blurring of these boundaries has brought new issues into the domain of international trade negotiations in a bid to ensure that policies and regulations are favourable towards the trading of compaks in the international market-place. By focusing upon the role of salient epistemic communities, Drake and Nicolaidis have been able to gauge the degree to which the actors involved in the recent round of trade negotiations have been able to reach consensus on the application of international principles for trade in services.[5] Their analysis concluded that the State actors who would become signatories to any new trade regime in the services domain would primarily reflect their 'material interests' (Drake and Nicolaidis 1992: 100), that is, the interests of private sector companies located within the boundaries of domestic (or regional) markets.

Recent expressions of instability in the international network governance regime for information and communication services have been ascribed to the emergence of a new market access regime. Cowhey and Aronson (1993: 184), for example, have suggested that this is a regime in which there is a relatively widespread consensus—at least among the major public and private actors—that 'the best guarantee of efficient competition is using trade or investment, including alliances, to introduce robust foreign competitors in all major economic centers' (Cowhey and Aronson 1993: 184). In this kind of regime 'trade and investment policies alone cannot suffice; complementary government policies also are needed to nurture industries' (Cowhey and Aronson 1993: 184). A market access regime favouring the material interests of the information and communication network and service producers and users is recognized as requiring a complicated balancing of market liberalization measures *and* measures to ensure that the use of networks and services generates an acceptable economic return to their producers.

The material or economic interests of the stakeholders in the electronic information and communication services market are substantial. Any emergent network governance regime is likely to be unsustainable unless it produces long-term benefits for the influential actors who participate in its creation and maintenance, and possibly over the death of earlier regimes. Within the framework of our middle-range perspective on technical and institutional change, the key issues concern the factors influencing the design of a new regime *and* those affecting the capabilities and actions of actors in the producer, consumer, and policy-making communities. The outcome of the regime formation processes that coincide with technical change are not determined in a straight-forward way by the power exercised by dominant industry actors or by the power of State actors. The shifting alignments among technology and service producers and users create the possibility that new incentives will be put in place for the production and consumption of electronic information services on the part of the many, rather than the few.

The following sections inquire into the extent to which the exclusionary effects of past network governance regimes are likely to be alleviated by ongoing negotiations and the formation of a new regime. The central tension in this process is the responsiveness of public and private sector actors to the growing demand for greater access to previously closed markets and for a tightening of intellectual property protection mechanisms. In the mid-1980s, for example, the ITU moved to modify the international regulatory regime for telecommunication to bring it in line with the emerging requirements of an open market access regime (International Telecommunication Union 1988).[6] During the same period, the WIPO moved to consider how copyright protection could be extended to electronic databases and computer programs (World Intellectual Property Organization 1983). By early 1994, the Marrakesh Agreement on modifications to the GATT had resulted in the coverage of advanced telecommunication services and the trade-related aspects of intellectual property (General Agreement on Tariffs and Trade 1994), and a new institution, the WTO, had emerged as a contender for the governance of information and communication technology and service markets. Based upon our research on the careers of these key institutions we highlight the contradictions, uncertainties, and the degrees of freedom available to the public and private sector actors as they seek to establish the policy and regulatory regime that will shape the technical design of the information and communication networks and services in the coming decades.

PRISING OPEN INTERNATIONAL COMMUNICATION:
ABOLISHING MONOPOLIES

The history of the global telecommunication infrastructure is a history of monopoly supply mainly by State-owned public telecommunication operators. These operators planned and constructed networks within the boundaries of nation States and engaged in international co-ordination to ensure network interoperability as and when they deemed this appropriate. This chapter is not the place to recount the ITU's history as a forum designed to enable these co-ordination activities to be carried out by the State representatives of the public telecommunication operators.[7] Instead, our task is to highlight some of the reasons for the winds of change which have challenged the viability of this institution as an effective governance forum in the 1980s and 1990s.

Since the mid-1980s the ITU has been challenged to provide an institutional framework for resolving conflicts among disparate private and public sector organizations. Yet it is an institution where the authority to govern is vested in the assumption that the issues it must resolve are primarily technical. The liberalization of national telecommunication markets has meant that the terms and conditions of market access have become the arbitrators of the international service revenue streams accruing to telecommunication service suppliers. This is a substantial change from a time when revenues accrued to operators as a result of the private agreements among publicly owned operators.[8]

The structural and regulatory characteristics of any two or more national markets create an environment which affects the incentive structure for the supply and use of the international telecommunication substructure. A decade ago, State- or privately owned monopoly telecommunication operators provided the substructure for the origination and termination of international telecommunication voice and data traffic. They jointly organized a framework which determines the contributions by operators to the costs of supply. Today, in most countries around the world, new network and service suppliers are seeking to originate and terminate traffic in ways that bypass these traditional accounting mechanisms. Although still subject to licensing and regulatory conditions in each national market, telecommunication supply has been liberalized such that operators based in one national market can compete in multiple foreign markets.

The new market access regime is organized so that, in principle, international traffic originating on a network infrastructure owned by operator A in country X can be terminated in country Y on operator A's network or on a network owned by another operator in country Y. Such traffic can be routed through a network in country Z owned by operator A, B, or C on the way to its destination. The traffic can be processed by operator A, B, C, or D as value is added to the information content by service providers. The organization and administration of billing, maintenance, security, and management can be integrated with the provision of a service by operator A in the originating country, or it can be handled by any one of a number of other parties.

The number of potential owners or operators of the network and service components which handle international telecommunication traffic could rise exponentially.[9] The share of traffic and revenues accruing to any single network operator or service provider, in theory, should reflect its technological superiority, the variety and quality of its services, and the efficiency of its operation. The technical characteristics linking this 'network of networks' also should be neutral as to cost and revenue advantage to any individual supplier. These characteristics would prevail, however, only in a market in which uniformity with respect to *market* access had been achieved on a world-wide basis.

The theoretical world of multiple, equally endowed independent players in the international market-place does not exist. Instead, alliances among some of the largest players characterize the market and there has been a substantial amount of new entry by niche market players of all kinds where this has been permitted by national authorities.[10] The transmission, switching, and information-processing components of networks are becoming akin to commodities that can be mixed and matched according to user specifications. This commodification of telecommunication services involves the unbundling and separate pricing of network functionality that traditionally had been sold by the public telecommunication operators as relatively homogeneous packages. The practices of mixing and matching hardware and software platforms in the computing industry are analogous to current developments in the telecommunication industry. As in the software industry, the availability of unbundled packages of network functionality does not eliminate opportunities for designing technical characteristics that give unfair advantage to certain suppliers.

Issues of compatibility and the need for open standards and protocols to support international networks in the monopoly telecommunication supply environment were addressed only at the interfaces between national telecommunication networks. The ITU provided a forum for negotiations which kept pace with innovations in the underlying technologies. However, with the digitalization of networks from the 1970s onwards and the growing reliance on software and virtual networks, international compatibility problems have reached into the heart of national communication networks. Today, numerous suppliers seek to originate or terminate traffic on the network substructure located within national jurisdictions. Equipment, network, and service suppliers and users have advocated the need for technical standards and regulatory institutions to resolve their problems in gaining access to (and operating within) national markets. In general, they have not regarded the ITU as being especially effective in providing these capabilities because they argue, as a governance forum, it continues to be influenced by the players who predominated in the monopolistic telecommunication supply era.

Nevertheless, the liberalization of telecommunication markets has not been characterized by a simple linear trajectory of institutional change in the governance regime. For example, new opportunities for open market access are not always welcomed by the would-be competitors and market liberalization can bring benefits which, after a certain point, may begin to be regarded as disadvantages (Noam 1992, 1994c). From the point of view of some network suppliers, too much liberalization and improved market access conditions can bring the risk that private investor strategies and commitments will be jeopardized. Too little liberalization and the presence of non-transparent conditions for market access, on the other hand, can leave the markets of the dominant public telecommunication operators relatively untouched by the entry of new competitors. Foreign entrants into domestic telecommunication markets have quite openly sought the protection of national regulatory authorities to ensure the long-term payback of their investments; and the larger the investment, the stronger this sentiment has often been (Mansell and Credé 1995).

In addition, larger business telecommunication users have often stressed the need for improved co-ordination among international network operators in order to support services which need to be provided on an international basis. When these users have been the most vocal champions of market liberalization measures and of greater differentiation of the services available on the market, they have also argued that there is a need for a degree of market co-ordination which, in turn, reduces incentives for competition (Mansell 1993*b*). Thus, there are tensions and contradictions in the views of even the largest suppliers and users in the telecommunication market-place as to an appropriate balance between co-ordination and competition. Some companies advocate uniformity in infrastructure services which they then can package and sell on to other users, or use themselves; while others seek quality and quantity differentiation throughout the network substructure and services market.

This is the market environment in which international network governance must be negotiated. The ITU has been characterized as 'a "control centre" for the provision, maintenance, operation, and co-ordination of international telecommunications' (Savage 1989: 14). Historically, it has not been concerned with the organization of a competitive market-place. The actors who participated in its activities were the State representatives of national public telecommunication operators. However, by 1993 ITU membership had been extended to some private operators of telecommunication networks. The ITU members are bound by the terms and conditions of the International Telecommunication Regulations which set out a broad framework for the co-ordination of international telecommunication services (International Telecommunication Union 1988). These cover the framework for sharing the revenues generated by international networks and the technical standards that are recommended for implementation by equipment suppliers, network operators, and service providers. The ITU also manages radio frequency spectrum allocations for civil applications including satellite, mobile radio, and broadcast services.

The challenge to the ITU has been to update its regulations, recommendations, and its decision-making procedures to reflect an international market-place populated by competing telecommunication network and service suppliers. The delegations which have pushed hardest for reform, including the USA and the UK, introduced competition in their domestic markets relatively early as compared to their counterparts in other countries.[11] Many of these countries are home to companies with substantial material interest in global equipment and information and communication service markets. The ITU's procedures for standards-making, radio spectrum allocation, the formulation of technical recommendations, etc., were considered by those countries that were encouraging competition in the domestic supply of infrastructure and/or services to be creating and maintaining exclusionary boundaries around other national territories. The result was that a closed club of public telecommunication operators was continuing to reap substantial monopoly profits. In some cases, surplus revenues contributed to the general revenues of the State, but the benefits of this arrangement were deemed to be outweighed by the inefficiency and lack of innovativeness of the State-owned monopolies. The ability of these monopoly operators to respond to the needs of customers was challenged within the ITU by potential competitors participating in the few State-led delegations that welcomed them (Bruce *et al.* 1986; Hills 1986), and by representatives of telecommunication user organizations.

The resulting reorganization of the ITU has been visible in three main areas. First, membership has been extended to corporate actors. Second, a business advisory forum comprising the chief executives of interested firms who advise the ITU secretariat has been created. And, third, the Secretary General of the ITU is now advised by a World Telecommunications Advisory Council comprised of senior executives representing major telecommunication companies with global interests (International Telecommunication Union 1993). These steps have begun to accommodate private producer and large telecommunication user interests within the boundaries of what was previously a closed forum for the governance of international telecommunications. However, the reorganization initiatives have brought the vulnerability of this governance institution to the surface. If it cannot gain the allegiance of the telecommunication equipment and service suppliers, there is a risk that the standardization and regulatory initiatives needed to achieve co-ordination will shift to other regional institutions. In regional fora, private sector firms may expect their interests to be more easily represented and this has already begun to occur as Richard Hawkins has suggested in Chapter 6.[12] By establishing closer relationships with private sector actors, the ITU also risks the possibility that divergent private sector interests will make it difficult to achieve consensus with respect to the common network interface standards and regulations that enable the relatively smooth interoperability of global 'networks of networks'.

The Organization for Economic Co-operation and Development (OECD) has also contributed to the destabilizing forces which have affected the career of the ITU. For example, the OECD held a meeting at the end of 1985 to attract ministerial attention to the importance of information and communication policy. Telecommunication services were discussed in the context of trade in services—at the time, an anathema to those who regarded international telecommunication supply as a bilateral relationship between State-owned operating companies or divisions of government. How could these organizations be engaged in trade? The OECD secretariat worked with delegation representatives to test the applicability of GATT concepts to tele-communication services. Discussions focused on how concepts such as most-favoured-nation treatment might apply and how arrangements for market access could be established (General Agreement on Tariffs and Trade 1989). This work helped to create momentum for the eventual inclusion of advanced telecommunication services in the general framework agreement on trade in services that emerged with the conclusion of the GATT Uruguay Round of Negotiations at the end of 1993.

The expansion of the number of ITU member delegations represent-ing both public and private interests is also proving to be a destabilizing force. There have been difficulties, for instance, in allocating radio spectrum for mobile radio service applications using satellite tech-nologies. The portion of the radio spectrum allocated for these services cannot accommodate all those organizations and consortia that seek to launch new services. The Federal Communications Commission in the USA has issued licences for operators of global services which could pre-empt European and other regional plans once they occupy the radio spectrum and this development has helped to push telecommunication services more centrally onto the trade in services agenda.[13]

European fears that companies based in the USA might dominate the satellite mobile radio service market have helped to exacerbate differ-ences in the positions of policy-makers from regional trading blocs. The negotiations within the ITU have also become characterized by the increasing predominance of discussions of the commercial interests of corporate actors in radio spectrum allocation (Sung 1992). These disputes over commercial and public interests in access to the radio spectrum in the context of mobile radio services are not the first time that political and economic concerns have come to the fore in the negotiation of the allocation of this resource. The long and often acrimonious debate over the 'first come first served' policy in the allocation of orbital slots for satellite communication in the 1970s created divisions among the political and economic interests of industrialized and developing countries (Smythe 1972, 1987). However, in the 1990s, the ITU must contend directly within its own decision-making apparatus with lobbying among private companies. Yet, despite having opened the door to private sector participation, the ITU has no mandate to engage in debate on the economic or political implications of issues involving spectrum allocation, or any other aspect of telecommunication that may have a bearing on trade policy.

The need for redesign of the network governance regime for telecommunications has been attributed to technical innovations, that is, the digitalization and computerization of networks, which have appeared to challenge the ITU's authority and its members' capabilities with respect to non-technical issues. However, it is the non-technical issues and the uneven process of market liberalization that have perhaps most substantially undermined the ITU's governance capabilities. In effect, the technical and socio-economic issues are now so visibly and explicitly intertwined that the argument that they can be kept separate in the practice of governance is becoming difficult to sustain. Soroos (1982: 67), for example, has argued that the ITU represented the 'prototype of rational policy making in which decisions were normally made on technical grounds', but empirical research is now available which demonstrates that this argument is unsustainable (Hawkins 1992).

Consensus-building with respect to ostensibly technical issues, for example, the resale and use of circuit capacity or procedures for establishing revenue-sharing arrangement among telecommunication operators, has become more difficult as companies join national ITU delegations. Motorola, for example, has representatives in several national delegations including Canada, France, and Australia as well as the USA and this is becoming a common practice for globally operating companies. Disparate corporate strategic orientations to different local and regional markets can create disputes over the desirability of open market access that are difficult to resolve in the absence of any trade dispute settlement mechanism within a forum such as the ITU.

The redesign of the network governance regime for telecommunication is a reflection of influential private (and some public) sector actors' concern that their material interests not be jeopardized by a failure to address controversies over technical issues that have taken on local, national, regional, and global dimensions. Furthermore, the convergence of all forms of electronic communication networks with the production of the information content must also be taken into consideration, interests that the ITU has no mandate to address. The remit of the ITU may be inadequate to cope with the sea changes in the technical capabilities of today's information highways. However, its capabilities are being bypassed largely for economic and political reasons by telecommunication network operators, equipment manufacturers, and multinational users as they turn to the dispute settlement mechanisms that they expect to be applied by the WTO.

PROTECTING INFORMATION MARKETS: STRENGTHENING 'MONOPOLIES'

The changes in the domain of the network substructure have led many observers to call for the establishment of a new regulatory paradigm which has competitive, rather than monopoly, forces at its core. For example, Paul Huber, who has closely followed changes in the USA and in global telecommunication markets argues that, 'we are moving toward a *new regulatory paradigm*, one of open market entry rather than exclusive franchise, of diversification and integration rather than quarantine, of competition rather than price regulation, of plenty rather than scarcity (emphasis added) (Huber 1993).

In Japan, the Ministry of Posts and Telecommunications is seeking to stimulate the expansion of networks throughout Japan and hopes to attract foreign investors to build new high-capacity infrastructure networks. In Britain substantial foreign investment is already present and more is promised in the development of broadband cable networks. In the USA, most foreign investors in the domestic infrastructure market are still subject to restrictions which stand uncomfortably beside that country's argument that all countries should bring down remaining barriers to inward investment and foreign supply of telecommunication infrastructure and services.

This new regulatory paradigm is mainly concerned with infrastructure and the carriage of signals—the substructure—although many of the major telecommunication infrastructure suppliers argue that they should be permitted to supply all electronic information and commun- ication services including broadcast entertainment services. Convergent network systems and multimedia services are transcending distinctions between technical modes of electronic service delivery and their information content. The contenders in national and global markets favour greater competition because, as Cruise-O'Brien and Helleiner (1980) have argued, trade does not follow the flag, it follows the communication system.

A focus on infrastructure or *carriage* policy and regulation, however, overlooks a potential contradiction within the newly emerging network governance regime. Multimedia services bring together additional players with an active interest in the future shape of this governance regime. Broadcasters, publishers, and the film producers, telecommun- ication equipment and network operators, and computing hardware and software producers, are entering alliances which have numerous permutations. Their concerns with the production and consumption of electronic information give them a material interest in ensuring that the new network governance regime is designed to create monopolies in information through a systematic tightening of the intellectual property- rights regime.[14]

On the one hand, open access to information and communication services is important for innovation and many suppliers acknowledge that network distribution channels should be open and accessible. On the other hand, this brings the danger that users will access information without adequately recompensing its producers. Restrictions on access and use of electronic information services therefore are deemed to be necessary. Information is characterized by attributes associated with public goods; producers have little incentive to engage in production of innovative services unless their endeavours generate appropriable and adequate returns (see Hirshleiffer 1973; Melody 1981). The production of information is characterized by high initial costs and indivisibilities, it is not destroyed by use, and it is difficult to assess the value of information unless it is accessed (see Boulding 1966; Lamberton *et al.* 1986; and Stigler 1961). The costs of excluding people from using electronic information services except under authorized conditions are high and involve technical means such as encryption and security devices as well as a variety of other enforcement mechanisms (Tang 1995).

The viability of the existing intellectual property-rights regime and alternative enforcement measures appropriate to electronic information falls within the remit of the WIPO and, since 1994, the WTO. The key problem in the development of a governance regime and institutions with the capacity to manage conflicts among the interests of the producers and users of information products is to resolve the tension between the benefits inherent in the effective use of information once it has been produced, and the need to create sufficiently strong incentives for producing information in the first instance (Foray 1994). As in the case of the network substructure regime, in the domain of information content, an international co-ordinating regime is needed both to ensure that electronic information is accessible and responsive to a wide range of social and economic needs as well as to protect the competitiveness of companies and rights in intellectual works.[15]

The WIPO and the UN Educational, Scientific and Cultural Organization (UNESCO) administer conventions applicable to the information products that are finding their way into global electronic networks. The WIPO and UNESCO administer conventions which establish standards for copyright for the protection of literary or artistic works.[16] The need for a degree of protection of intellectual creations generally is not disputed. Rather, it is the design of an effective governance regime and consensus on minimum standards of protection that are the subject of controversy. We do not present a detailed history of the institutions involved in copyright protection in this chapter, nor are we concerned with the details of legal interpretations of available protections for electronic information. The recent history of regime formation in this area and its culmination in the design of a new institutional framework under the auspices of GATT is the main focus since it provides a contrast with the network substructure regime. In contrast to the emphasis on competition within the substructure regime, the overriding goal of the content regime is the strengthening of, rather than the erosion of, monopolistic supply.

The proliferation of electronic databases and computer software programs—in fact, all electronic information and communication services accessible via world-wide networks and on disc—has proved difficult to accommodate within the terms of existing intellectual property-rights conventions. In 1983, a committee of experts on the legal protection of computer software within the WIPO began to consider measures that would address the need to monitor the commercial exchange of software and other forms of electronic information products. In this forum, debate centred on whether a new draft treaty for the protection of computer software should be introduced. The committee took the view that it would be premature to take any stand on this issue (World Intellectual Property Organization 1983). By 1992, however, draft protocols for the protection of computer programs, databases, programme rental rights, and signals emitted by direct broadcast satellites had emerged (World Intellectual Property Organization 1992).[17]

The Berne Convention, the convention with the largest number of contracting parties (countries), has proven to be a difficult framework within which to accommodate advanced information and communication technologies and services. If, for example, a computerized music database with a catalogue of recordings allowing the user to select, listen, and obtain the history and the name of the recording artist, is downloaded onto a user's computer, there is no clear protection available to prevent the downloaded information from being reused for commercial purposes (Wired 1994).

The momentum toward the redesign of international institutions capable of protecting private interests in electronic information began to increase in the USA in the 1970s when the Reagan administration embarked on a programme to strengthen international protection of intellectual property. The cornerstone of this initiative was to be the inclusion of negotiations on intellectual property-rights issues within the GATT framework and this was achieved in 1986 (General Agreement on Tariffs and Trade 1986). The US delegation to GATT argued that some aspects of intellectual property rights involved trade issues. Software and audio-visual programme piracy had become of major concern to US-based producers (Nimmer 1992).[18] The move to the GATT forum was expected to offer a possibility for the design of a new code of conduct that would be enforceable under the GATT dispute settlement mechanism. Some developing countries initially opposed proposals in this area (as they had in the context of the liberalization of network services) arguing that there was no independent empirical evidence of the economic harm claimed by the USA and attributed to the existing regime (McDowell 1994). Midway through the process of negotiation and assessment the position of these countries was portrayed by one observer in the following terms.

A GATT agreement on (Trade Related Intellectual Property Rights) TRIPS, . . . and certainly a multilateral framework on Services, are being negotiated to provide enhanced protection and rights for Transnational Corporations. Even elementary countervailing obligations, such as disciplines on corporate entities and their practices, have been ruled out of court. . . . They should make sure that corporate policies and practices are explicitly covered and disciplines on governments are matched by disciplines on private operators in the market. (Raghavan 1990: 290, 305)

Representatives of these countries argued that low-cost access to information products is a prerequisite for technological innovation and that stronger protection mechanisms would simply create disincentives for innovation (Yusuf and von Hase 1992). Nevertheless, in 1991 the UN Conference for Trade and Development (UNCTAD) took a favourable view of the development of stronger enforcement mechanisms for intellectual property rights in a trade context:

The new framework, if finally adopted, will make it harder to follow the 'catching up' practices used in the past, and will entail better devised strategies for technical development. The challenge for developing countries is great, as is the responsibility of the international community to ensure that the new rules do not deepen the present economic and technological asymmetries. The search for new mechanisms of international cooperation to ensure that all countries have access to the fruits of technological progress should thus rank high on the agenda of the international community for years to come. (UNCTAD 1991: 194)

The view that stronger protection of intellectual property is beneficial because it increases the flow of technology and knowledge and stimulates investment by facilitating licensing agreements continued, nevertheless, to lack independent empirical documentation. As Cottier (1991: 391) suggested, 'by and large, the matter is still largely one of political and ideological debate rather than in-depth analysis and field studies.' Developing countries began to shift their position as negotiators came to believe that higher standards of copyright protection could reinforce their retaliatory powers in trade disputes. The WIPO forum continued to be perceived to be making too little progress toward establishing a new basis for the protection of works produced by, or embedded in, information and communication technologies.

Unevenly established national and regional standards for the protection of intellectual property as well as major differences in conditions of enforcement had presented continuing problems for the international content governance regime. The USA, for example, did not accede to the Berne Convention until 1988 as its domestic copyright protection legislation was regarded as being more stringent than the protection offered by international conventions.[19] It finally did so under pressure from information and communication service suppliers in the USA who argued that failure to accede to the Berne Convention would render the

US-based companies vulnerable to piracy. The USA might also find it difficult to press its case for a new international regime in the GATT context if it did not become a signatory to a convention which underpinned the existing regime (Motyka 1992).

The European Union had agreed a directive by 1991 that accorded protection to computer programs under copyright law as literary works (Council of the European Communities 1991). A draft directive on the protection of computerized databases was issued by the European Commission in October 1993 but, as of spring 1995, this was still under discussion.[20] The proposed directive was formulated on the assumption that the imbalance in the level of investment in database creation between the member States of the European Union, and between the Union and the world's largest database-producing country, that is, the USA, must be redressed by giving stronger protection to the creators of such works.

The enforcement capabilities of the existing intellectual property-rights regime was also subject to criticism by those seeking greater protection. The international conventions provide for settlement of disputes by the International Court of Justice, but WIPO-administered conventions had been criticized for their inadequacy with respect to enforcement. By 1990, WIPO had formulated a new draft treaty covering the settlement of disputes (World Intellectual Property Organization 1990), but by this time the bandwagon effects of pressures to strengthen the world intellectual property rights regime had strengthened. In 1991, stronger enforcement was high on the agenda of the major information and communication service-producing countries (Cottier 1992). A G7 Economic Summit pronounced that the GATT was the appropriate forum for the resolution of intellectual property-rights trade-related disputes. 'The principal requirement is to move forward urgently [on] intellectual property, where clear and enforceable rules and obligations to protect all property rights are necessary to encourage investment and the spread of technology' (Group of Seven 1991). With the conclusion of the Uruguay Round of trade negotiations at the end of 1993, a new trade-related aspects of intellectual property-rights agreement was in place—administered by the new WTO.

This new agreement embraces copyright, trademarks, patents, protection of trade secrets and undisclosed information, the subject-matter of broadcasts, and the reproduction of phonogrammes. Computer programs are defined as 'compilations of data or other material, whether in machine readable or other form, which by reason of selection or arrangement of their contents constitute intellectual creations' (General Agreement on Tariffs and Trade 1994: Annexe 1C, Art. 10). However, the coverage of information and communication services was not fully resolved. Electronic publications and derivatives of computerized databases, for example, fall in a grey area of intellectual creations not clearly addressed by any existing convention or multilateral agreement.

Not only did the new agreement fail to catch up with technical developments in information and communication technologies and services, it also excluded a crucial issue that reflected strong tensions between the interests of private entities in maintaining exclusive (monopolistic) rights to information and the wider diffusion of information products which can contribute to innovation. Article 6 of the agreement on trade-related aspects of intellectual property states that: 'nothing in this Agreement shall be used to address the issue of *exhaustion* of intellectual property rights' (emphasis added) (General Agreement on Tariffs and Trade 1994: Annexe 1C, Art. 6).

The exhaustion of rights is a complex issue and it is critical to whether the owner of intellectual property rights in one country can use these rights to stop another country from importing the same goods or services from the same title-holder but via a different country (Yusuf and von Hase 1992). Uniformity imposed by a GATT regime on the exhaustion of territorial rights could undermine the export-based economic development strategies of developing countries and those undergoing a transition to market economies in central and eastern Europe. Basically, rights holders can invoke their exclusive rights to protect their home markets from parallel imports. Territorial exhaustion has an effect such that if a developing country, for example, relies on foreign technology or an imported information product for its domestic development, it could then be restricted in its choice of export markets. Under the new agreement, countries continue to be able to operate their own systems of national (regional) exhaustion of rights depending on their circumstances (Reinbothe and Howard 1991), and so they retain a degree of flexibility within the new albeit stricter regime for electronic information content.

As a result of the agreement, enforcement measures are to be strengthened and anti-competitive practices are to be subject to consultation between governments or to referral to the dispute settlement mechanism under GATT. The new electronic content regime is intended to encourage the acquisition of capabilities to protect and enforce intellectual property rights by countries so as to 'contribute to the promotion of technological innovation and to the transfer and dissemination of technology, to the mutual advantage of producers and users of technological knowledge and in a manner conducive to social and economic welfare, and to a balance of rights and obligations' (General Agreement on Tariffs and Trade 1994: Annexe 1C, Art. 7).

INTERNATIONAL GOVERNANCE: CONVERGENT CARRIAGE AND CONTENT REGIMES?

In a world characterized by complex network substructures linking information and communication service producers and users into networks that do not stop for inspection at the boundaries of the nation State, it is not surprising that a new institutional regime should emerge both for the substructure and for content. As Dunning has suggested, there is likely to be a continuing process of 'recasting of international supervisory or control mechanisms, e.g. GATT, to take account of specific attributes of global production' (Dunning 1992: 327). From the perspectives of the players with major stakes in these global markets, neither the ITU nor the WIPO had shown sufficient signs of forging a consensus on how the global market should be co-ordinated or how the players should compete within it.

Both of these institutions were perceived, at least by some of the public and private stakeholders, as being all too ready to be concerned with social, distributional, and development-related issues. In the case of the ITU, many delegations continued to advocate regulations and recommendations that would protect their monopoly operators and nationally based equipment suppliers. In the case of WIPO, delegations equivocated on the need for a stronger regime since a more forgiving one might enhance their access to electronic information enabling them to strengthen their capabilities as producers—rather than simply as users—of information and communication services.

The economic and social development-related issues that are of concern to these delegations are not excluded from consideration in the GATT. Provisions for the least developed countries are explicitly present in both the agreement on advanced telecommunication services and on the trade-related aspects of intellectual property rights. However, the delegations within the GATT context represent a more balanced mix of government and private sector representatives than is the case in the older issue-specific governance institutions. WIPO, for example, has been described as being 'checkmated by the Third World . . . Part of that attitude comes from the sense that one cannot accomplish the trade-offs that are possible across sectors . . . within WIPO' (American Society of International Law 1990: 266–7) and the US delegation particularly was seeking a more hospitable forum for negotiation.

The GATT framework is facilitating the co-ordination of market access rules for the builders and users of the network substructure and the standards for the protection of intellectual property rights. Like the ITU and the WIPO, however, it does not do so neutrally. The capabilities of the ITU and the WIPO governance regimes appear to have ebbed and they may be approaching the end of their effective life-cycles as the arbitrators of the fortunes of producers and users of the international network substructure and globally traded electronic information and communication services. There is little doubt that market access and competition are on the ascendancy as far as the development of the distribution channels for electronic information and communication services are concerned; or that a tightening of the intellectual property regime will continue with respect to information services.[21]

The major public and private players in the information and communication technology and services markets have the capability to design a new regime that is consistent with their economic and political interests. And, as US Vice-President Gore (1993) has argued, the shift to a market access regime within the GATT/WTO framework will help to 'clear from the road the wreckage of outdated regulations and allow for a free-flowing traffic of ideas' (Gore 1993). Some observers argue that firms will underinvest in R&D in the absence of adequate protection of their intellectual property and that a global scale of operation is necessary to generate financial returns substantial enough to build the next generation of information highways and to launch new electronic information services. However, there may be lessons to learn about the effects of the newly emerging regime from the experience of patenting in other industrial sectors. For example, in 1989 only one per cent of existing patents were held by nationals of developing countries and patenting activity in the less developed countries was dominated by non-residents (OECD 1989). Penrose (1951: 220) has suggested that, 'non-industrialized countries and countries in the early stages of industrialization gain nothing from granting foreign patents since they themselves do little, if any patenting abroad. These countries receive nothing for the price they pay for the use of foreign inventions or for the monopoly they grant to foreign patentees.' Despite the dearth of independent empirical information with respect to the impact of the existing copyright and other forms of protection applicable to electronic information services, strengthened protection is to be extended on a world-wide basis. The quasi-monopolistic rights that this protection confers on title-holders implies increases in the costs to users accessing and using the electronic information that will circulate through future information highways.

The rhetoric surrounding the proliferation of information highways and the growth of multimedia markets has succeeded in shifting the concerns of public and private actors away from issues of 'territorially bounded places' to use John Ruggie's (1993) term. Ruggie has argued that the global economy of the 1990s is a 'decentred yet integrated space of flows' of information which exists alongside national economies. Distinctions between the two are becoming increasingly blurred and this is resulting in increasing complexity for the design of institutions that can provide negotiating fora that address both the economic and the social issues associated with the development and diffusion of new information and communication technologies and services.

Suppliers and users of electronic services increasingly operate as if they were virtually sovereignty-free and their actions are said to have lost a strong connection with sovereign authorities (Rosenau 1992). Yet the new WTO regime for the network substructure and electronic information content has been designed as if sovereign authorities were still firmly in place. The larger suppliers and users and their trade associations try to ensure that market liberalization measures, and those for the protection of intellectual property reflect their economic interests in both their home or primary markets and those foreign markets they are penetrating around the world. The new regime is deeply embedded in the sovereign-State apparatus and this suggests that some consideration will be given to public interest issues which extend beyond the material interests of these companies.

Luard has argued that the ITU and the WIPO, historically, were concerned mainly with problems of co-ordination that arose out of economic efficiency and wealth-generation considerations on the part of the major players, but that they were also concerned with wealth distribution and equity issues. Thus, 'the widening gap between rich nations and poor, combined with the narrowing physical distance which makes the discrepancies more visible, brings more urgent demands for international redistribution of wealth. . . . It will be a struggle not only about the *means* of international government. It will concern the ends it should serve' (Luard 1977: 322–3). In the new GATT/WTO framework, delegations may be less concerned with distributional and equity issues as representatives of the private sector promote their material interests in wealth-generation.

The declining fortunes of institutions like the ITU and the WIPO as viable network governance regimes have been attributed to their inappropriateness in the face of technical innovations in information and communication technologies and to the 'decline of distance in the modern world' (Luard 1990: 180). This conclusion rests on a functionalist analysis of the life-cycle of international governance institutions in which they cohere or disintegrate on the basis of whether they fulfil a given set of functions (Mitrany 1966). In this view, governance institutions do not learn or acquire new capabilities.

Yet another perspective assesses the viability of regimes in terms of the norms, principles and values which guide and facilitate decision-making (Taylor 1990). Questions about the life-cycles of institutions within these regimes focus on whether they allow decisions about new problems to be made quickly and encourage actors to accept short-term costs in order to increase long-term benefits (Krasner 1991). From this perspective, both the ITU and WIPO would be expected to decline as both have experienced difficulties in adapting the speed of their decision-making apparatus to the pace of technical change. They also have failed, respectively, to convince the majority of their members to open their markets to the supply of telecommunication infrastructure, and to strengthen and enforce the protection of intellectual property.

CONCLUSION: DESIGNING GOVERNANCE INSTITUTIONS

The foregoing perspectives on the births and deaths of international governance institutions do not adequately take into account the strength of the tensions (the dialectic) in the formation of new regimes. The durability of existing regimes and institutions such as the ITU and the WIPO is associated with the complexity of the co-ordination and competition arrangements that need to be negotiated. The sheer complexity of global information highways and the relative ease with which information and communication services can be appropriated will provide an impetus to learning processes and to the acquisition of new capabilities by the old and new institutions of governance.

Increasing differentiation and specialization characterizes the 'network of networks' of the 1990s. This creates the need for 'co-ordination between subsystems . . . and such problems require institutional bridges between the different subsystems' (Traxler and Unger 1994: 17). The GATT/WTO regime provides a new bridging mechanism that may be effective in resolving some of the conflicts between an open market access regime for the information and communication substructure and a restrictive monopolistic regime for the protection of electronic information. However, the capabilities of the older institutions, that is, the ITU for co-ordination of aspects of network design, and the WIPO for development of treaty instruments for the protection of intellectual property, will not be easily replicated.

What scope does the new network governance regime offer to meet the needs of socio-economic communities who have little or no access to communication networks and few opportunities to contribute to, or to benefit from, the world's growing stock of electronic information products? Our focus on institutional design and capabilities shows that the degrees of freedom available to actors involved in governance cannot be read directly from the formal texts and constitutions of the governance institutions or from the exercise of power by individual public or private actors at any single point in time. Like the life-cycles of technologies, the life-cycles of institutions are not characterized by a clear progression from birth to death. They display complex patterns of negotiated boundaries, changes in the capabilities of their participants, and in the power actors are able to exercise within them.

The international network governance regime is not simply reactive to the progression of a technological juggernaut. In the case of the ITU, its viability has been challenged by the uneven liberalization of national telecommunication markets and by the ascendancy of a market access regime; the conclusion has been that a new regime is needed. A similar conclusion has been reached in the case of the WIPO in the intellectual property-rights domain. One response has been the creation of a new institution, the WTO. However, the new institution's capabilities do not match the complexity of the technical system that it must govern. The market access regime represented by the WTO is unlikely to encourage actions sufficient to redress emerging gaps between the information-rich and the information-poor and these distributional questions will create the need for further creative regime-building.

The coming period of instability in international governance in this area is likely to see the maintenance of parallel regimes, that is, the GATT/WTO and the ITU/WIPO regimes, growing complexity in their institutional boundaries and capabilities, and continuing uncertainty in the growth prospects for the larger globally operating suppliers and users in spite of the creation of a new network governance regime. The way in which contradictions between the old and the new network governance regimes are resolved will affect opportunities for improving access to the network substructure and the degrees of freedom available to balance interests in the protection of rights to electronic information and in the wider diffusion of such information.

NOTES

1. The network substructure is comprised of network switching, routing, management and transmission technologies, see Ch. 6.
2. Finkelstein (1991: 13) defines governance as 'all overlapping categories of functions performed internationally, among them: information creation and exchange; formulation and promulgation of principles and promoting consensual knowledge affecting the general international order, regional orders and particular issues on the international agenda'. Regimes are characterized by conflict between individual agents' preferences or between those expressed by collective groups (Commons 1932).
3. The ITU and WIPO are specialized agencies of the UN. WIPO was established in 1967 and became operational in 1970. The ITU was established as the International Telegraphic Union in 1865. The WTO was established with the Marrakesh Agreement in 1994 and is intended to facilitate the

implementation, administration, and operation of multilateral and pluri-lateral trade agreements including the GATT, 1947 and subsequent amendments.

4. In 1994 the OECD (1994c: 1) argued that 'modern communications infrastructures, and their corresponding services, have already transformed production and distribution processes in manufacturing, have become an integral part of service industries, and are a key factor in international trade and the globalization of the world economy.'

5. An epistemic community is a network of knowledge-based experts which plays a role in articulating the cause-and-effect relationships of complex problems, is instrumental in helping States to identify their interests, framing issues for collective debate, proposing specific policies, and identifying salient issues for negotiation (Haas 1992).

6. The Final Acts of the 1988 World Administrative Telegraph and Telephone Conference (WATTC) included 'special arrangements' which ensured access to competing telecommunication operators under certain conditions in markets which previously had been closed to competitive entry, see International Telecommunication Union (1988: Art 9). The ITU has also undergone a process of internal reorganization to enable private sector suppliers to engage more fully in its activities (International Telecommunication Union 1989, 1993).

7. The early history of the ITU is described in Codding (1972), Codding and Rutkowski (1982), and Goldberg (1985).

8. In countries such as the USA and parts of Canada, the operators were privately owned regulated monopolies. International telecommunication traffic on the world's public switched telecommunication networks grew in 1992 by approximately 12% and growth is expected to continue to be strong.

9. For a more detailed discussion of the characteristics of the full competition model in the telecommunication sector, see Mansell (1993a).

10. e.g., alliances have been struck between MCI and British Telecom; France Telecom and Deutsche Telekom, and consortia such as AT&T's Worldsource. Companies in the computer software industry, for example, Microsoft, are planning to launch global information networks that will compete with services offered by the traditional telecommunication companies. In March 1995 a software company in Israel had developed software to enable the Internet to be used for voice telephony, albeit at reduced quality.

11. Competition in infrastructure supply in the USA began in the late 1960s and in the UK in the mid-1980s.

12. Co-ordination issues are addressed in the context of the economics of compatibility standards for the equipment and networks, see David and Foray (1994b), David and Greenstein (1990), and Hawkins *et al.* (1995).

13. In 1992 the FCC notified the ITU of its spectrum requirements for low earth orbiting satellites. At the World Administrative Radio Conference (WARC) in 1992, L-band spectrum was allocated to mobile satellite services. In late 1993, the FCC indicated its intention to license suppliers observing that the provision of global services by companies in the USA might enhance their presence in world markets.

14. Convergence is evidenced by alliances such as those between AT&T, NCR,

and McCaw Cellular; US West, Time Warner, and Mercury Communications; Telewest, Reuters, and London Radio News; News International's purchase of Delphi Internet Services (on-line service); and Paramount Communications' acquisition of MacMillan Inc. in the USA. However, many of these alliances have proven to be short-lived.

15. Co-ordination with respect to the rules governing the market for electronic information exchange have not been studied in-depth although the implications of intellectual property rights regimes for innovation and competitiveness have been the focus of much work, see e.g. David (1993), Meessen (1987), OECD (1989), and Office of Technology Assessment (1986).

16. The Berne Convention of 9 Sept. 1886, revised Paris 24 July 1971 and amended 2 Oct. 1979 covers 'literary and artistic works' (Contracting Parties 1886). Disputes can be brought before the International Court of Justice. UNESCO administers the Universal Copyright Convention of 1952 (Contracting Parties 1952), extending protection to authors and other copyright proprietors of literary, scientific, and artistic works, including writings, musical, dramatic and cinematographic works, and paintings, engravings, and sculpture. Disputes can be brought before the International Court of Justice. WIPO administers the Rome Convention for the Protection of Performers, Producers of Phonograms and Broadcasting Organizations of 1961 (jointly with UNESCO), the Geneva Convention for the Protection of Producers of Phonograms against Unauthorized Duplication of their Phonograms of 1971, the Satellite Convention of 1974 and the 1989 Treaty on Intellectual Property in Respect of Integrated Circuits.

17. Protection was extended to photography in 1908, cinematography and musical works in 1948, and broadcasting in 1974.

18. In 1988 firms in the USA received $US 8 bn. in intellectual property royalties while paying firms in other countries only $US 1.25 bn. in royalties; firms in the USA claimed to have lost some $US 43 bn. as a result of piracy and counterfeiting (American Society of International Law 1990).

19. For a comprehensive review of the situation in the USA, see Lehman (1994). In Nov. 1994 a court decision overturned existing legal provisions for the protection of databases.

20. The UK publishing industry has criticized the draft, arguing that the protection it extends is insufficient as provisions are made for 'unauthorized extraction' which some publishers argue will undermine the intent of copyright protection (Commission of the European Communities 1993c).

21. In February 1995 the USA and Japan targeted China for its abrogation of the duty of the State to introduce adequate protection of CD-ROMs and other forms of electronic information and communication services, thereby delaying China's possible accession to the GATT. Although press announcements claimed a resolution of this dispute in March 1995, it is unclear whether effective enforcement mechanisms will be put in place by the Chinese government.

8

The Politics of Information and Communication Technologies

ROGER SILVERSTONE AND ROBIN MANSELL

Throughout this book we have been discussing the politics of information and communication technology. This politics, however, is not only or even most significantly a politics of nations or parties. It is a politics of the innovation process as such, a politics engaged in by participants—individuals as well as institutions—at every point in the production and consumption of new information and communication technologies and services. This politics is marked by great inequalities: inequalities generated by the power of regulatory institutions and multinational companies, and the uneven capacity of consumers to enter and manage a rapidly changing technological space. It is also marked by great uncertainties and discontinuities: uncertainties and discontinuities generated as a result of the shifting tectonic plates of a global information and communication economy, as well as the opacity, fragility, and unpredictability of consumer behaviour. It is, finally, a politics deeply embedded not just within the institutions that design and distribute technologies and services, but within the technology itself, as software products and information networks both prescribe and proscribe, configuring suppliers and users, containing and constraining behaviour, and embodying in their algorithms and their gateways both the normative and the seductive.

A 'BODY' POLITICS

The politics and economics of innovation is, in a Foucauldian sense, a 'body' politics. It is a politics in which technical design is inescapably intertwined with human capabilities. At issue is the capacity of actors at every level not to gain control of the innovation process as a whole, for as we have argued, as information systems become ever more complex that increasingly seems to be an impossibility, but to gain control of their little (or big) bit of it. The struggles for control are fought out on the world stage as well as in the front room, behind the closed doors of national and international regulators as well as in the showrooms of the high street, in the design of telecommunication networks and customized software as well as in their implementation and use. Information and communication technologies are not just material objects. They have created and sustained an encompassing symbolic space, both vital and virtual, in which power is claimed, exercised, and disputed.

We have approached these complex issues from a number of different disciplinary perspectives. We have focused on the process of innovation: on its contradictions and on its tensions as well as on its social, political, and economic determinants and consequences. At every point we have stressed that information and communication technologies have to be understood as social, that technical potential has to be realized, that determinations are never fully determining, and that the interactions of production and consumption, of State and market, of local and global interests, produce degrees of uncertainty and conflict whose results can neither simply be read off from an analysis of the 'technologic' of research and development nor be assumed from the blandishments of politicians or marketeers. We have pointed out that there are biases, unevennesses, and inequalities in these processes: and that though inevitably some actors have more power than others and yet still others are excluded almost entirely from the game, that the power that is held is always contested and that the loci of control are constantly shifting as institutions and individuals manœuvre to gain maximum leverage on electronic spaces and markets, both in public and in private.

THE MIDDLE RANGE

In framing our approach we have done two things. We have defined—in broad terms—what Robert Merton has called a middle-range theory, and we have operationalized that theory through attention to the relationships between design and capabilities. However, the reader will have noted that we have not attempted to construct a middle-range theory in the singular. Our perspectives, as well as the range of problems that we address, are too diverse for that to become possible or even desirable. Indeed, we are mindful of Merton's own caution, when he says:

The middle-range *orientation* involves *the specification of ignorance.* Rather than pretend to knowledge where it is in fact absent, it expressly recognizes what must still be learned in order to lay the foundation for still more knowledge. It does not assume itself to be equal to the task of providing theoretical solutions to all the urgent practical problems of the day but addresses itself to those problems that might now be clarified in the light of available knowledge (emphasis added). (Merton 1968, 68–9).

Our theories are akin to what Merton dubs an *orientation* and through them we have been pursuing both the commonalities as well as the specificities of structure and agency in the technical and institutional innovation process. They are, further, grounded in a set of arguments and assumptions that have one of their sources, perversely perhaps, in the grand theorizing of the sociologist Anthony Giddens. Structuration theory in its various formulations has, despite itself, the pretensions to universality which both we and Merton would reject. Indeed, it is not unique in its approach. Yet it provides a ground base for offering a sensitive sociology, economics, and politics of innovation, enabling attention to the dialectics of structure and agency, power and resistance, institutionalization and creativity, which define and drive technical innovation.

THE VIRTUAL STRUCTURE

Giddens's immersion into the classic and fundamental dilemma of sociology offers a number of insights into the tensions at the heart of social and economic life which must have a bearing on our understanding of the ways in which information and communication technologies are becoming engrained into the fabric of institutions and everyday life. His terms—structure and agency—are difficult and abstract, and a number of his critics have had difficulties particularly with the former. Giddens's reply is both instructive and curiously of direct relevance to our arguments here. Structure, he suggests, 'is what gives form and shape to social life, but it is not itself that form and shape . . . it exists only in a virtual way' (Giddens 1989: 256).

Structure exists only in its manifestation in action, but actions themselves are dependent on rules and formulae, of varying degrees of specificity and visibility, which enable and constrain, legitimate and sanction. At the interface of structure and agency meanings are produced and values created and enforced and so too are the material artefacts of the information and communication industries. The capacity to do so: to participate, to engage, to influence, is a function of power, of the power of 'structure' to limit the activities of agents, and of the differential power of agents to 'structure' their own socio-economic and technical environments.

These insights are relevant to our own project in a number of important ways. They enable us to enquire into the particular dynamics and dialectics of innovation at the point at which new technologies are introduced into complex organizations and institutions as well as at any other point in their conception and use. However, they also offer a much more fundamental route into the specificity of information and communication technologies as in turn, in the late twentieth century, they are fundamental to our capacity both to construct meaning and to create and communicate value. The identification of social structure as 'virtual' is significant, for it is the virtuality of socio-economic structure created by the emergence of information networks and by software and information of various kinds and the capacity of these structures too to facilitate and constrain, that we can begin to recognize their *essentially* social as well as technical nature.

This is perhaps the key point, and its ramifications are pursued in the various contributions to this book. It is to show in what ways and with what consequences information and communication technologies are both interdependent with, and interpenetrate, the socio-economic world.

STANDARDS AND CAPABILITIES

Standards provide, in may ways, the starting-point, for it is in the establishment of common procedures and protocols, common both technically and in the widest sense also linguistically, and common to all those with the desire and resources to gain access to the system that emerges, that communicative networks become possible. Flawed though it is, both in conception and realization, the design of information and communication technologies, is both the product and producer of human capabilities. Nothing, no commerce, no communication, is possible without the capacity and competence to connect. As Robin Mansell has argued 'electronic commerce is feasible only to the extent that an appropriate network infrastructure or "substructure" is in place' and that this substructure is itself the product of, and dependent upon, a prior structuring of software and informational capabilities. The networks of course have no significance until they are used, and their use is conditioned not just by their endogenous characteristics, but by external constraints of a more obvious (though still submerged) socio-economic kind. And there is a further point buried within the etymology of the term commerce: that it refers not just to commercial transactions but to 'intercourse in the affairs of life' (OED).

The process of standards-setting is, therefore, in many ways exemplary. This process is both social and intensely political as well as economic. Richard Hawkins has pointed to the imperfections and inequalities of access to the fora within which technical standards of various kinds are to be agreed, and also to the containing politics as well as the costs of access, itself conditioned by changes in technology as well as in industrial alliances, and vulnerable to the uncertainties of competing and changing interests and agenda. This process has a certain logic, but it is a situated logic produced at the interface of the negotiating process itself, and it is not to be defined by any abstract or singular model of rationality. The emerging standards are the product then of a complex interrelationship of necessarily interrelated standards-setting bodies as well as the activities of governmental and commercial agents and agencies. The results are an inevitable compromise, expressive of the tension between the technical imperative to produce a compatible and interconnectable system and the inevitable distortions of a negotiating machinery in which regional, national, and local political and economic interests confront and undermine one another. Here is a supposedly technical procedure being reread as a social one, but as Hawkins himself points out, the results 'embody many political and commercial biases and [yet] still perform more or less adequately as technical elements in the network'. This 'more or less adequate' process is instructive and it has a wider relevance. It signals the human capacity to work with even technically created imperfections, and even more than that. As Robin Mansell points out in her discussions of information networks, in such imperfections lies the potential for suppliers and users to gain the degrees of freedom necessary to create a measure of competitive advantage.

Software, as Paul Quintas argues, is another powerful example of the interpenetration and interdependence of the technical and the socio-economic in the world of information and communication technologies. Neither the process of their development nor their implementation and use is without its contradictions. The 'faking' of the order that is presumed to be required as a precondition for the production of order, the inability to provide formal procedures for the effective delivery of the perfect software machine, the politics of control between producer and user, and the tension between inherent flexibility and path dependency are all elements of a complex and conflicting world of situated rationalities. The politics of software development is a politics, on the one hand, of the supposed need to standardize and regulate the production process; and, on the other, of the competing demands of users and producers to determine the character of what increasingly amounts to the operating language of organizations and institutions as well as of machines. The software industry is constantly changing, but the loci of control have perceptibly shifted towards those who control the standardizing and packaging of commodified products. This has inevitable consequences for the way in which users use software, but more than that. In the structures of code, and masked by the veneer of infinite flexibility, are the kind of constraints that Michel Foucault would have gladly recognized. The structures of knowledge embedded in the formal procedures and facilitating virtual environments which advanced software products release, are neither ideologically neutral nor infinitely malleable.

The message is embedded in the medium—at the point of delivery, but most especially through time, as social structures encrust around the procedures built into the codes and output of even the most sensitively customized software application. In the same way that institutions move uncertainly and in a sometimes disorderly way from charismatic to rational and then to traditional forms of domination and legitimation (Weber 1978), so too does software; from its creation in the under-disciplined craftshops of the software houses; to its implementation as a principle mechanism of apparently rational order and efficiency in machine or institution as the virtual machine; to the heavy hand of electronic concrete, denying the possibilities of change by the sheer weight and rigidity of overmaintained codes. Software, like other technologies, has its career and its life-span. To follow this career is to follow the cultures that form around and become dependent upon it, and it is to be able to explore, especially in its implementation, the struggles for control as both individual and institutional users seek to match the potential inscribed within it to what they understand as their own needs, and through that matching, to master it.

Conflicts between producers and consumers which result in the accommodation of the new to the demands and pressures of the user are not always especially visible in the production of software, especially since as Paul Quintas points out, a high proportion of software development takes place in-house. More visible and more significant perhaps both for our understanding of the nature of the transaction between producers and consumers as well as its consequences for the receiver and user, is the case of the household and the family. Roger Silverstone and Leslie Haddon's discussion of the design–domestication interface raises a number of points relevant to the specificity of information and communication technologies, to their commodification, and to the implications of technical and media change both for the politics of the household and for the wider politics of consumption.

REFLEXIVITY, RATIONALITY, AND UBIQUITY

Information and communication technologies, systems, and services are essentially reflexive. That reflexivity in part emerges in Roger Silverstone and Leslie Haddon's description of such technologies as doubly articulated in the social: that such machines, their hardware and their software, are the focus of meaning construction at the same time as they enable it. The liberation of communication from its dependence on human or mechanical carriage begun with the telegraph has continued, and gathered increasing pace, such that neither space nor time are any longer significant constraints, although they remain significant as dimensions of the communication process. From quantity to quality: from speed to instantaneity. Information no longer seems to travel. It just seems to arrive. Yet a complex substructure is intertwined with these technologies and with what, when, and how they enable us to communicate. The technologies and communicative action have become inseparable. David Harvey (1989) has identified this reflexive capacity of information and communication technologies as a key to an understanding of the transition from modernity to postmodernity: to the paradoxical world of globalization and fragmentation which such reach and speed occasions.

Yet such globalizing analysis leaves relatively untouched the experiences of those whose lives are conducted both with or without technology and who have to survive the rigours of daily life. The structuring of our symbolic environments which these technologies uniquely undertake is an activity which still has to be managed: accepted, rejected, transcended, or transformed. It follows that the domestication of information and communication technologies is an activity which involves content as well as object, both software and hardware. It involves, together, the medium, the message, and the machine. But it also extends beyond the home. To equate these technologies' ubiquity with their omnipotence is, therefore, to mistake the complexities and subtleties of individual and organizational behaviour. Our argument has identified the centrality of design for an understanding of the emergence and specificity of information and communication technologies only in order to problematize it. The discussion of the design–domestication interface is intended to signal the role of the user in the process of innovation, but it also signals the presence and the power of the situated rationalities of home and household, and the many other environments, which are grounded in the multiple experiences of daily life, which are themselves the product of class, ethnicity, or geography, and which change through time as households move through their life-cycle. Both Jan Pahl (1989) and Lydia Morris (1990, 103–22) have in their different ways drawn attention to the capacity of households to define for themselves their own economic rationality, articulated through the control, management, and budgeting of family finances and resources; and Jonathan Parry and Maurice Bloch (1989) have shown how that rationality is the product of distinct domestic cultures, defined alongside the economic rationality of public and institutional life but neither obeying the same allocative rules nor expressing the same defining values.

This account of the situated economics of the household is anticipated in the various attempts to extend institutional and evolutionary economics to embrace the dialectic of technical innovation. Robin Mansell's discussion of that thread of economic analysis which takes its start from, among others, the work of Thorstein Veblen (1919), draws attention to the dynamic nature of economic action, both dependent upon and in turn facilitating the capabilities of actors to exercise a degree of freedom in defining for themselves the terms within which economic, as well as social and political, action, is pursued. This approach shifts attention away from the exogenous to the endogenous determinants of such action, and once again requires us to frame our understanding of the power of information and communication technologies in terms of their structurable as well as their structuring potential. In this context, Mansell's extension of concepts of learning and producer–user inter-action developed by economists including Dosi *et al.* (1994), Lundvall (1992*c*), and Nelson (1994) frames her inquiries into the deeper process of technical and institutional change in firms and governance organiza-tions; processes which have proven so illusive within the formalism of economic and political theory. An individual's or an institution's capabilities in this context are a product therefore not just of the availability of the resources released by information and communication technology but of the skills and competencies, as well as the desires and values, of those who both produce and consume.

CONSUMPTION AND COMMODIFICATION

Information and communication technologies, perhaps uniquely then, are invented at the point of consumption. Once again the double move in our analysis directs us both to the ways in which our focus on these technologies requires us to problematize the relationship between production and consumption, and, *per contra*, our focus on the relations of production and consumption leads us to rethink the character of the technologies themselves. The boundary between production and consumption is blurred not just by the fact that at the institutional level consumers are also in turn producers, but by the fact that in software, quite specifically, and in the fragmentation of markets and taste cultures more generally (Featherstone 1991), the consumer or business user is embodied in the design of the object. This embodiment is itself contradictory and never straightforwardly the result of a direct interface between producers and users. The boundary is, finally, blurred by the constant interaction of the user with the meanings and significance of the soft- and hardware of the information society. The end-user programmer is alive and well at computer terminals and tele-communication handsets both in public and private spaces. Nowhere is this more visible, of course, than in the Internet and on the World Wide Web.

To focus on the relationship between production and consumption in the information and communication economy is to focus, necessarily now, on the process of commodification. We have already seen, following Paul Quintas's account of the changing economics of software, how its packaging and mass production is inevitably both affecting the capacity to design user-specific programs (and users, like travellers on British Rail, are now hailed not as clients or passengers but as customers) and also leading to increasing problems for the protection of intellectual property. But it is Rohan Samarajiva who confronts the commodity head on, tracing both the extension of commodification throughout the information and communication economy in which users are progressively being caught in the increasingly fine mesh of the world-wide commodity web, and its reverse face: the hidden surface of economic (and political) surveillance that accompanies the informatisation of consumption.

Samarajiva's argument is that both in the commercial environment of product choice and mass marketing and in the supposedly less rapacious environment of the public communication utility, practices and networks are being created—designed—in such a way as to facilitate the exercise of increasingly subtle and pervasive forms of surveillance. To consume is to declare not just one's hand but one's body. To pick up the phone is to step out of the shadows of private space into a virtual public world of infinite information control and intrusive access. In the first context, that of the commercial transaction, Samarajiva points to two sets of practices that both separately and together provide an infrastructure for the increasing capacity of those who attempt to manage markets to gain information about, and control over, their customers. The *compak* builds on the efforts, through market research and marketing to define, increasingly precisely, the characteristics of consumers and to gain maximum leverage on their future consumer behaviour. It involves the creation of an integrated package of goods and services which enables the quite precise identification of distinct types of individual consumers and which then seeks to lock them into a relationship that, in turn, enables the continuing channelling of now carefully targeted messages and products. Buying a new car or a software package increasingly involves the consumer in a set of service arrangements which both generates information about his or her patterns of consumption and use and creates a dependency which it is intended should lead to further sales. As a result the supposedly greater and greater choices being offered to the consumer mask a consumer culture in which, in fact, individual consumers are being offered less and less, because those who sell no longer need to waste valuable breath (and resources) in trying to persuade Jo and Joanne Doe to buy something they might not want.

Transaction-Generated Information (TGI) is, in turn, the tip of an iceberg of private/public information produced by specific information-gathering processes but especially by the collocation of data gathered from different sources. TGI can be generated in a number of different ways, and though inevitably still quite imprecise, it is invisible and involves no additional participation from the consumer. It is at its most insidious, Samarajiva argues, when it is generated by monopolistic firms and when it is based on records of participation in an electronic information and communication environment.

Both the compak and TGI are illustrative of the increasing power and intrusiveness of the process of commodification and of the capacity of the market to manage private decisions and private actions. The surveillance capacities increasingly being built into the telecommunication network, however, have a different significance, since they involve, for example in calling line identification, the capacity of either a private monopolistic firm or the State to intrude into, and to control, public electronic space.

None of these activities is, of course, unproblematic. The capacity totally to control the market is likely to be chimerical despite significant advances in information about consumer behaviour. And, in the case of the public communication utility, forms of information-gathering are increasingly being subject themselves to public scrutiny and the possibility of challenge. Herein lie some of the indeterminacies in the system. As Samarajiva suggests, 'unlike the private domains in which mechanisms of consumer surveillance are being installed, these public spaces afford greater possibilities for intervention by citizens and public advocacy groups.' At issue is not only the capacity of market or market-related information to be used against the consumer but the undermining of the serendipity that defines the character of communication in public spaces, and an increasing capacity to intrude into the private spaces of individuals and organizations. We can speculate therefore on the capacity of citizens' groups as well as those desirous of open and unpoliced systems for market access, to mount an oppositional politics to contain or even restrain these developments.

Two quite different things are signalled by this discussion. The first is the problem, raised separately by Roger Silverstone and Leslie Haddon and by Richard Hawkins, of finding the user. The second is the role of public policy in managing the societal implications of the information and communication revolution addressed by Robin Mansell. We postpone discussion on the latter until later on in this concluding chapter, but it is worth making a few observations about the first now. Silverstone and Haddon identify the importance of finding and catching the consumer as a stage in their model of the innovation process. This task was made more critical by the perceptual uncertainty of demand for new information and communication technologies and services and the great difficulties for those involved in all stages of the design process to imagine who would use their new products and how. If information and communication technologies really are invented in consumption, then so too, it seems, are the consumers. Richard Hawkins takes this requirement to find the user more literally, pointing to the shift in the balance between network provider and telecommunication user with the emergence of the mobile phone. The new mobility in public space which this occasions forces the network operator to locate the user, rather than the other way around. Consumers, then, in this new world of private mobilization, are both literally and metaphorically, increasingly nomadic, moving in and out of product spaces in the same way as they move in and out of media and telecommunication range.

ACCESS AND ACTION

There are those of course who never are in range. Any assumption of the homogeneity and ubiquity of the information and communication revolution is seriously flawed. At the same time the presumption that there are (and maybe always will be) a society of information-rich and information-poor, while easy to assert, needs careful analysis if it is to be useful in the determination of policy. Robin Mansell has offered indicative evidence that even after three generations of electronic communication and information provision, large areas of the world have telephone and television penetration rates in single figures. This profound inequality in the global diffusion of even basic information and communication technologies is of considerable significance, such that discussions either of the global village or even of a more measured attempt to generate a kind of technological leapfrog from the technologically primitive to the technologically advanced seems like so much hot air.

Even in advanced industrial societies such as the USA and the UK, distinctions and divisions between those who have access to the present generation of information and communication technologies are significant enough to warrant caution and to be sceptical about the grander claims for an information society. Access to computers among school-age children in the US is profoundly skewed both at school, but especially at home, against Blacks and Hispanics. Even more profound are the inequalities as measured by a comparison between those earning less than $US 20,000 and those earning more than $75,000.[1] Similar differences in home computer ownership are evident in the UK, and while in no way as profound there are still manifest inequalities in ownership of telephones in the UK (MacKay, 1995). But these figures, significant though they are, mask other factors and issues.

Crude figures of ownership, especially, say little about both the quality of the technology that is owned and the way in which it is, or is not, used. Nor do they enable us to see whether and how such disparities and biases are being recreated or subverted through the politics of innovation in information and communication technologies. Old and ailing media technologies providing poor reception and low-level services are much more likely to be found in the homes of the poorer and more disadvantaged in our society. Those who live in such homes too are likely to have fewer developed skills in their advanced use. New machines and services will not only take much longer to reach such people, but if and when they do, they may not be supported by the kinds of literacy and competence necessary to take advantage of what is on offer. The availability of services, both in the sphere of information and entertainment, is further constrained by the intrusion of the market and the consequent weakening of commitments to public service in broadcasting and to universal access in telecommunications. While societies as a whole become progressively more dependent on electronic media and information and communication services, then those who find access, for financial or other reasons, increasingly difficult or just simply impossible will become progressively marginalized and excluded. Isolation can take many forms. In an information age, where technologies themselves are reasonably seen as the instruments of social and economic inclusion, then non-participation in a culture defined and articulated through electronic networks of one kind or another, will have profound cultural as well, as political and economic consequences.

The progressive interdependence of socio-economic and electronic structures, their shared virtuality, is therefore double-edged. It enables as well as constrains, includes as well as excludes. The capacity to participate in electronic culture and to move within electronic space is in a real sense, and increasingly, a basic requirement for engagement in the society in which we live. That participation is a precondition for agency, and for effective action, either in compliance or opposition. The politics of the electronic information and communication age requires, as both Mansell and Samarajiva in their different ways argue, the capacity to develop effective human communicative capabilities. Those capabilities, in turn, depend on the degrees of freedom that the subject within a technologically mediated culture can effectively mobilize. It may be that the nature of that subjectivity and the requirements for its effective realization are changing. Our activities in the private spaces of everyday life as well as in the public spaces of work are defined and constrained by our location within a superordinate electronic space. While we have argued against determinist positions of various kinds, stressing above all else the essentially social, and therefore human, nature of technology, we are not denying the reality of change nor the enormous challenges that such change generates. On the contrary, the possibilities for effective political action, as well as the capacity to generate coherent policy, depend on the need, together, to develop an understanding of the mechanics and mechanisms of communication and information, the need to develop an understanding of the contrariness of the information and communication revolution, and the need not to be blinded by the revolution's pervasive rhetorics.

TOWARDS POLICY IN A DIALECTIC OF HUMAN ACTION

Our discussions have focused principally on the uncertainties of the interrelationships between the technical and socio-economic orders. We have done this in order to draw out the deeper processes that inform the complex relationships between information and communication technology producers and users in their changing environments. In the home, in the workplace, and in the governance institutions, both public and private policies are inextricably part of this landscape. Robin Mansell's analyses of the strategies of private firms and of the life-cycles of international governance regimes show how critical the decisions of actors in these institutionalized frameworks can be to the design of technologies and our capabilities to use them.

Our insistence on considering these policy environments—their structural features and the agency of human actors—as neither determining nor entirely indeterminate, offers a framework within which policy-making itself can be seen to be designed and articulated not just in the formal 'corridors of power' but in the everyday interventions and experiences of information and communication technology producers and users.

The literature on 'policy' or 'strategy', in the academic world and in the trade and policy domains, shows a burgeoning of expressions of interest and concern about the 'cultural', 'social', 'political', and 'economic' impact of the information and communication technologies revolution. Should, for example, the media-producing industries be regulated, and how; should the shackles be lifted from former monopoly players in the telecommunication field; what legislation or regulations should apply to emergent monopolists; should standards be mandated— when and how should this be accomplished; should it be permissible for information concerning private individuals to be disclosed and what technological means should be encouraged to achieve such protection as is deemed necessary; should the State legislate against, or in favour of, the importation of distant entertainment programming or software; should we seek (within the presently understood boundaries of the State) a greater degree of competence in the production of the software and hardware of the information and communication revolution; and should there be greater expenditure by the State or by private investors on the skills of consumption through new education and training programmes? Should resources urgently be pumped into the construction of the substructure and the content of electronic information and communication? Is there a need for specific policy interventions or should the market run its course? These are just some of the questions and issues that are discussed in the boardrooms of firms and in public policy institutions.

These questions are all part of the wider politics of information and communication technologies which we have addressed in this book. The answers become embedded in the information and communication technologies and services as they are produced and consumed. They are changeable through time and they must be the subject of continuous inquiry, inquiry into the principal contradictions and the dialectic of the processes of socio-economic and technical change.

The cumulative histories of policies, their careers through the births and deaths of informal and formal public and private institutions, and their interdependence with the design and capabilities of the technical world of information and communication technologies are the proper subject matter of socio-economic analyses of the longer- and medium-term trajectories of technical change. Our goal has been to illustrate the value of middle-range theories in 'opening the black box' on the vicissitudes and deeper processes of this innovation process. Our theoretical and methodological approaches are consistent with the aim of moving forward towards a workable understanding of technical and institutional change; an understanding that can be employed in both formal and informal policy settings, to seek improvements in the quality of our socio-economic and technically mediated lives.

We have suggested that the information and communication technologies of the late twentieth century are designed, but we have shown that this process is itself problematic. The prevailing perception is that individual consumers, firms, and policy-making institutions are confronted by, and need to adjust to, the exigencies of 'the greatest technological juggernaut that ever rolled'. Analysis grounded in a dialectic of human agency and structure brings to light, in contrast, the constraints to, and the possibilities of, expressions of power by and through the development and implementation of these, ever more pervasively present, technologies.

Wiener (1950: 25, 28) argued that 'society can only be understood through a study of the messages and the communication facilities which belong to it; and that the further development of these messages and communication facilities, messages between man and machines, between machines and man, and between machine and machine, are destined to play an ever-increasing part. . . . Thus communication and control belong to the essence of man's inner life even as they belong to his life in society.' We have argued that the institutionalization of the production and consumption of symbolic or informational content and the technical means of communication are critical features of today's socio-economic and political order. In this book, we have offered glimpses of how this perspective illuminates some of the urgent problems in science, technology, economic, and other realms of policy-making. We refer the reader to our publications throughout the course of the Programme on Information and Communication Technologies for examples of the way this perspective has informed analysis of issues on the policy agenda in the UK, the USA, Europe, and internationally (Science Policy Research Unit 1995). The continuing challenge is to frame the insights of our inquiries and those of others in a way that moves forward our understandings and conceptions of innovation in disparate technical and organizational settings. Our contribution to middle-range theories of innovation in information and communication technologies seeks to uncover those situations and actions that can make a difference. In so doing, it offers provocations and opportunities to those who seek a less divisive and harmful socio-economic and technical world.

NOTES

1. US Census 1993 figures, cited in *Newsweek* (1995).

BIBLIOGRAPHY

Abernathy, W. J., and Utterback, J. M. (1975) 'A Dynamic Model of Process and Product Innovation', *Omega*, 3(6): 639–56.

—— —— (1978) 'Patterns of Industrial Innovation', *Technology Review*, 80 (June/July), 2–29.

Adelson, A. (1993) 'Software Sounds the Alarm on Cellular Theft', *New York Times*, 18 July.

Adler, E., and Haas, P. M. (1992) 'Conclusion: Epistemic Communities, World Order, and the Creation of a Reflective Research Programme', *International Organization*, 46(1): 367–90.

Aglietta, M. (1979) *The Theory of Capitalist Regulation*, London: New Left Books.

Altman, I. (1975) *The Environment and Social Behaviour: Privacy, Personal Space, Territory and Crowding*, Monterey, Calif.: Brooks/Cole.

Amendola, M., and Gaffard, J.-L. (1994) 'Out of Equilibrium Process of Change', paper presented at the EUNETIC Conference, Evolutionary Economics of Technical Change: Assessment of Results and New Frontiers, European Parliament, Strasbourg, 6–8 Oct.

American Society of International Law (1990) 'What's Going on in International Property Law?', in American Society of International Law (ed.), *84th Annual Meeting of the American Society of International Law*, Washington DC, 256–77.

Aoki, M. (1990) 'The Participatory Generation of Information Rents and the Theory of the Firm', in M. Aoki, B. Gustafsson, and O. Williamson (eds.), *The Firm as a Nexus of Treaties*, London: Sage, 26–52.

Arthur, W. B. (1989) 'Competing Technologies, Increasing Returns and Lock-In by Historical Events', *Economic Journal*, 99 (Mar.), 116–31.

—— (1990) 'Positive Feedbacks in the Economy', *Scientific American*, (Feb.), 80–5.

—— (1993) 'On Designing Economic Agents that Behave like Human Agents', *Evolutionary Economics*, 3: 1–22.

Baggott, R. (1986) 'By Voluntary Agreement: The Politics of Instrument Selection', *Public Administration*, 64 (Spring), 51–67.

Baldwin, D. A. (1980) 'Interdependence and Power: A Conceptual Analysis', *International Organization*, 34(4): 471–506.

Banham, R. (1960) *Theory and Design in the First Machine Age*, London: Architectural Press.

Barré, R. (1994) 'Interactions between International Innovation Networks within MNFs and National Innovation Systems: "Residual" or New Paradigm?', paper presented at the EUNETIC Conference, Evolutionary Economics of Technical Change: Assessment of Results and New Frontiers, European Parliament, Strasbourg, 6–8 Oct.

Barry, A. (1990) 'Technical Harmonisation as a Political Project', in G. Locksley (ed.), *The Single European Market and the Information and Communication Technologies*, London: Belhaven, 111–20.

Bateson, G. (1979) *Mind and Nature: A Necessary Unity*, New York: Bantam Books.

Baudrillard, J. (1981) *For a Critique of the Political Economy of the Sign*, St Louis: Telos Press.

Baudrillard, J. (1988) *Selected Writings*, ed. M. Poster, Cambridge: Polity Press.

Bauer, J. M. (1994) 'The Emergence of Global Networks in Telecommunications: Transcending National Regulation and Market Constraints', *Journal of Economic Issues*, 28(2): 391–402.

Bauman, Z. (1988) *Freedom*, Minneapolis: University of Minnesota Press.

—— Cantell, T., and Pederson, P. P. (1992) 'Modernity, Postmodernity and Ethics—An Interview with Zygmunt Bauman', *Telos*, 93: 133–44.

Bell, D. (1973) *The Coming of Post-Industrial Society: A Venture in Social Forecasting*, New York: Basic Books.

Benedikt, M. (1992) *Cyberspace: First Steps*, Cambridge, Mass: MIT Press.

Beniger, J. R. (1986) *The Control Revolution: Technological and Economic Origins of the Information Society*, Cambridge, Mass: Harvard University Press.

—— (1993) 'Communication—Embrace the Subject, Not the Field', *Journal of Communication*, 43(3): 18–25.

Bernstein, C., and Woodward, B. (1974) *All the President's Men*, New York: Simon & Schuster.

Besen, S. M. (1990) 'The European Telecommunications Standards Institute: A Preliminary Analysis', *Telecommunications Policy*, 14(6): 521–30.

—— and Saloner, G. (1988) 'The Economics of Telecommunications Standards', in R. W. Crandall and K. Flamm (eds.), *Changing the Rules: Technological Change, International Competition, and Regulation in Communications*, Washington, DC: Brookings Institution, 177–220.

Bhaskar, R. (1989) *The Possibility of Naturalism Theory*, 2nd edn., Hassocks: Harvester Press.

Bijker, W., and Law, J. (1992) *Shaping Technology/Building Society: Studies in Sociotechnical Change*, Cambridge, Mass: MIT Press.

—— Hughes, T. P., and Pinch, T. (eds.) (1987) *The Social Construction of Technical Systems*, Cambridge, Mass: MIT Press.

Bloomfield, B. P., and Vurdubakis, T. (1994) 'Boundary Disputes: Negotiating the Boundary between the Technical and the Social in the Development of IT Systems', *Information Technology and People*, 7(1): 9–24.

—— Coombs, R., and Owen, J. (1994) 'The Social Construction of Information Systems—The Implications for Management Control', in R. Mansell (ed.) *The Management of Information and Communication Technologies: Emerging Patterns of Control*, London: Aslib, 143–57.

Boehm, B. W. (1981) *Software Engineering Economics*, Englewood Cliffs, NJ: Prentice-Hall.

—— (1988) 'A Spiral Model of Software Development and Enhancement', *IEEE Computer*, 21(5): 61–72.

Booch, G. (1994) 'Coming of Age in an Object-Oriented World', *IEEE Software*, 11(6): 33–41.

Boulding, K. E. (1966) 'The Economics of Knowledge and the Knowledge of Economics', *American Economic Review*, 56(2): 1–13.

Bourdieu, P. (1977) *Outline of a Theory of Practice*, Cambridge: Cambridge University Press.

—— (1984) *Distinction: A Social Critique of the Judgement of Taste*, London: Routledge & Kegan Paul.

Bressand, A., Distler, C., and Nicolaidis, K. (1989) 'Networks at the Heart of the Service Economy', in A. Bressand and K. Nicolaidis (eds.) *Strategic Trends in Services: An Inquiry into the Global Service Economy*, Grand Rapids, Mich.: Ballinger Publishers, 17–32.

Breyer, S. (1982) *Regulation and its Reform*, Cambridge, Mass: Harvard University Press.

Brooks (Jr.), F. P. (1986) 'No Silver Bullet: Essence and Accidents in Software Engineering', in *Proceedings of Information Processing '86*, North-Holland, Elsevier Science Publishers BV: Amsterdam, repr. in T. DeMarco and T. Lister (eds.) (1991), *Software State-of-the-Art: Selected Papers*, New York: Dorset House, 14–30.

Brown, J. S. (1991) 'Research that Reinvents the Corporation', *Harvard Business Review*, Jan.–Feb., 102–11.

Bruce, R. R., Cunard, J. P., and Director, M. D. (1986) *From Telecommunications to Electronic Services: A Global Spectrum of Definitions, Boundary Lines, and Structures*, London: Butterworth.

Bucciarelli, L. L. (1984) 'Reflective Practice in Engineering Design', *Design Studies*, 5(3): 185–90.

Buckminster-Fuller, R. (1971) 'Introduction', in V. Papanek (ed.), *Design for the Real World: Human Ecology and Social Change*, New York: Pantheon Books, pp. vii–xix.

Burns, R., Samarajiva, R., and Mukherjee, R. (1992) *Customer Information: Competitive and Privacy Implications*, Columbus, Oh.: National Regulatory Research Institute.

Business Week (1991) 'Can the US Stay Ahead in Software', *Business Week*, 11 Mar., 98–105.

Butler Cox Foundation (1990) 'Electronic Marketplaces', Butler Cox Foundation: London.

Cabinet Office, Information Technology Advisory Panel (1983) *Making a Business of Information*, London: HMSO.

Calhoun, G. (1992) *Wireless Access and the Local Telephone Network*, Norwood, Mass.: Artech House.

Callon, M. (1986) 'The Sociology of an Actor Network: The Case of the Electric Vehicle', in M. Callon, J. Law, and A. Rip (eds.), *Mapping the Dynamics of Science and Technology*, London: Macmillan, 19–34.

—— (1987) 'Society in the Making: The Study of Technology as a Tool for Sociological Analysis', in W. E. Bijker, T. P. Hughes, and T. J. Pinch (eds.), *The Social Construction of Technological Systems: New Directions in the Sociology and History of Technology*, Cambridge, Mass.: MIT Press, 83–103.

—— (1992) 'The Dynamics of Techno-Economic Networks', in R. Coombs, R. Saviotti, and V. Walsh (eds.), *Technological Change and Company Strategies: Economic and Sociological Perspectives*, New York: Harcourt Brace Javanovich, 72–102.

—— and Law, J. (1982) 'On Interests and Their Transformation: Enrolment and Counter-Enrolment', *Social Studies of Science*, 12: 615–25.

Campbell, C. (1987) *The Romantic Ethic and the Spirit of Modern Consumerism*, Oxford: Blackwell.

Campbell, C. (1991) 'Consumption: The New Wave of Research in the Humanities and Social Sciences', *Journal of Social Behaviour and Personality*, special issue, 6(6): 57–74.

Cargill, C. (1989) *Information Technology Standardization: Theory, Process, and Organizations*, Maynard, Mass.: Digital Press.

Carlsnaes, W. (1992) 'The Agency-Structure Problem in Foreign Policy Analysis', *International Studies Quarterly*, 36 (Sept.), 245–70.

Carmel, E. (1993) 'How Quality Fits into Package Development', *IEEE Software*, 10(5): 85–6.

—— (1994) 'Can the USA Stay Ahead in Packaged Software', Department of Management Working Paper WP-94-10, American University: Washington, DC.

—— and Becker, S. (1995) 'A Process Model for Packaged Software Development', *IEEE Transactions on Engineering Management*, 41(5), in press.

Carnavale, M. L., and Lopez, J. A. (1989) 'Making a Phone Call Might Mean Telling the World about You', *Wall Street Journal*, 28 Nov.

Castells, M. (1993) 'The Informational Economy and the New International Division of Labor', in M. Carnoy, M. Castells, S. S. Cohen, and F. H. Cardoso (eds.), *The New Global Economy in the Information Age*, University Park, Pa.: Pennsylvania State University Press, 15–43.

Cawson, A., Haddon, L., and Miles, I. (1990) 'Successful Consumer IT: A Literature Review and Synthesis', a Report prepared for British Telecom Research and Technology Division, RT 5447, Brighton: University of Sussex, Feb.

—— Miles, I. and Haddon, L. (1995) *The Shape of Things to Consume*, Falmer: Falmer Press.

Central Statistical Office (1994) 'Annual Abstract of Statistics 1994', Central Statistical Office, London: HMSO.

Cheal, D. (1988) *The Gift Economy*, London: Routledge.

Ciborra, C. U. (1994) 'Market Support Systems: Theory and Practice', in G. Pogorel (ed.), *Global Telecommunications Strategies and Technological Changes*, Amsterdam: North-Holland, 97–110.

Coase, R. H. (1937) 'The Nature of the Firm', *Economica*, 4 (Nov.), 386–405.

—— (1993a) 'The Nature of the Firm: Influence', in O. E. Williamson and S. G. Winter (eds.), *The Nature of the Firm: Origins, Evolution and Development*, New York/London: Oxford University Press, 61–74.

—— (1993b) '1991 Nobel Lecture: The Institutional Structure of Production', in O. E. Williamson and S. G. Winter (eds.), *The Nature of the Firm: Origins, Evolution, and Development*, New York/Oxford: Oxford University Press, 227–35.

Cockburn, C. (1992) 'The Circuit of Technology: Gender, Identity and Power', in R. Silverstone and E. Hirsch (eds.), *Consuming Technologies: Media and Information in Domestic Spaces*, London: Routledge, 32–47.

—— and First-Dilic, R. (eds.) (1994) *Bringing Technology Home: Gender and Technology in a Changing Europe*, Milton Keynes: Open University Press.

Codding, G. A. (1972) *The International Telecommunication Union: An Experiment in International Cooperation*, New York: Arno Press.

—— and Rutkowski, A. M. (1982) *The International Telecommunication Union in a Changing World*, Dedham, Mass: Artech House.

Cohen, S. (1985) *Visions of Social Control: Crime, Punishment, and Classification*, Cambridge: Polity Press.

Commission of the European Communities (1989) 'ESSI: The Case for a Major European Systems and Software Initiative', Commission of the European Communities, DGXIII, Brussels.

—— (1990) 'Perspectives for Advanced Communications in Europe PACE '90', i: Commission of the European Communities, Brussels, Dec.

—— (1992) 'Report on Multimedia', Interservice Group, Analysis of Industrial & Technological Strategies, DGXIII, Commission of the European Communities, Brussels.

—— (1993*a*) 'Research and Technology Development in Advanced Communications Technologies in Europe, RACE 1993', Commission of the European Communities: Brussels, Feb.

—— (1993*b*) 'Overview of the Japanese Electronic Information Services Market', Commission of the European Communities, Information Market Observatory, Brussels.

—— (1993*c*) 'Amended proposal for a Council Directive on the Legal Protection of Databases', COM(93) 464 final—SYN 393, Commission of the European Communities, Brussels.

Commons, J. R. (1932) *Institutional Economics: Its Place in Political Economy*, Madison, Wis: University of Wisconsin Press.

Comor, E. A. (1994) 'Introduction: The Global Political Economy of Communication and IPE', in E. A. Comor (ed.), *The Global Political Economy of Communication*, New York: St Martin's Press, 1–18.

Contracting Parties (1886) 'Berne Convention for the Protection of Literary and Artistic Works of September 8 1996, revised Paris, July 24 1971 and amended October 2 1979', Geneva: World Intellectual Property Organization.

—— (1952) 'Universal Copyright Convention', presented to Parliament by the Secretary of State for Foreign Affairs by Command of Her Majesty, Nov. 1957, London: HMSO (Treaty Series No. 66).

Cottier, T. (1991) 'The Prospects for Intellectual Property in GATT', *Common Market Law Review*, 28: 383–414.

—— (1992) 'Intellectual Property in International Trade Law and Policy: The GATT Connection', *Aussenwirschaft*, 47 (Jan.), 70–105.

Council of the European Communities (1991) 'Council Directive of 14 May 1991 on the Legal Protection of Computer Programs', Council of the European Communities, 91/250/EEC, 14 May.

Cowhey, P. (1990) 'The International Telecommunications Regime: The Political Roots of Regimes for High Technology', *International Organization*, 44(2): 169–99.

—— (1993) 'Domestic Institutions and the Credibility of International Commitments: Japan and the United States', *International Organization*, 47(2): 299–326.

—— and Aronson, J. (1993) 'A New Trade Order', *Foreign Affairs*, 72(1): 183–95.

Cox, R. W. (1981) 'Social Forces, States and World Orders: Beyond International Relations Theory', *Millennium: Journal of International Studies* 10(2): 125–55.

Cox, R. W. (1987) *Production Power and World Order: Social Forces in the Making of History*, New York: Columbia University Press.

Cruise-O'Brien, R., and Helleiner, G. K. (1980) 'The Political Economy of Information in a Changing International Economic Order', *International Organization*, 34(4): 445–70.

Cyert, R., and March, J. (1963) *A Behavioural Theory of the Firm*, Englewood Cliffs, NJ: Prentice-Hall.

Dalum, B., Johnson, B., and Lundvall, B.-A. (1993) 'Public Policy in the Learning Society', in B.-A. Lundvall (ed.), *National Systems of Innovation: Towards a Theory of Innovation and Interactive Learning*, London: Pinter, 296–317.

Dankbaar, B., and van Tulder, R. (1991) 'The Influence of Users in Standardization: The Case of MAP', MERIT WP 91–013 Maastricht Economic Research Institute on Innovation and Technology, University of Limburg, Maastricht.

Danziger, J., Dutton, W., Kling, R., and Kraemer, K. (1982) *Computers and Politics*, New York: Columbia University Press.

David, P. A. (1986) 'Understanding the Economics of QWERTY: The Necessity of History', in W. N. Parket (ed.), *Economic History and the Modern Economist*, Oxford: Blackwell, 30–49.

—— (1987) 'Some New Standards for the Economics of Standardization in the Information Age', in P. Dasgupta and P. Stoneman (eds.), *Economic Policy and Technological Performance*, Cambridge: Cambridge University Press, 198–239.

—— (1993) 'Intellectual Property Institutions and the Panda's Thumb: Patents, Copyrights, Trade Secrets in Economic Theory and History', in M. B. Wallerstein, M. E. Mogee, and R. A. Schoen (eds.), *Global Dimensions of Intellectual Property Rights in Science and Technology*, Washington, DC: National Academy Press, 19–61.

—— and Foray, D. (1994a) 'Accessing and Expanding the Science and Technology Knowledge-Base: A Conceptual Framework for Comparing National Profiles in Systems of Learning and Innovation', report prepared for the OECD Directorate of Science, Technology and Industry, DSTI/STP/TIP(94)4, Paris, 25 Apr.

—— and Foray, D. (1994b) 'Percolation Structures, Markov Random Fields and the Economics of EDI Standards Diffusion', in G. Pogorel (ed.), *Global Telecommunication Strategies and Technological Changes*, Amsterdam: North-Holland, 135–70.

—— and Greenstein, S. (1990) 'The Economics of Compatibility Standards: An Introduction to Recent Research', *Economics of Innovation and New Technology*, 1(1): 3–41.

—— and Steinmueller, W. E. (1990) 'The ISDN Bandwagon Is Coming, But Who Will Be There To Climb Aboard? Quandaries in the Economics of Data Communication Networks', *Economics of Innovation and New Technology*, 1(1): 43–62.

David Shepard Associates (1995) *The New Direct Marketing: How to Implement a Profit-Driven Database Marketing Strategy*, 2nd edn., Homewood, Ill.: Dow Jones-Irwin.

Davidow, W. H., and Malone, M. S. (1992) *The Virtual Corporation: Structuring and Revitalizing the Corporation for the 21st Century*, New York: Harper-Collins.

Davis, D. M. (1994) 'Going in Circles in Mass Communication: From Mass Audiences to Mass Producers, Illusions and Ambiguities in Human Roles and Intelligent TeleMediated Channels', paper presented at the 19th International Association for Mass Communication Research Conference, Seoul, Korea, July.

Dawe, A. (1979) 'Theories of Social Action', in T. Bottomore and R. Nisbet (eds.), *A History of Sociological Analysis*, London: Heinemann, 362–418.

de Sola Pool, I. (1977) *The Social Impact of the Telephone*, Cambridge, Mass.: MIT Press.

Diesing, P. (1971) *Patterns of Discovery in the Social Sciences*, Chicago: Aldine Atherton.

Dobuzinskis, L. (1992) 'Modernist and Postmodernist Metaphors of the Policy Process: Control and Stability vs. Chaos and Reflexive Understanding', *Policy Sciences*, 25(4): 355–80.

Donzelot, J. (1979) *The Policing of Families*, London: Hutchinson.

Dordick, H. (1981) *The Emerging Network Marketplace*, Norwood, NJ: Ablex Publishing.

Dosi, G. (1982) 'Technological Paradigms and Technological Trajectories: A Suggested Interpretation of the Determinants and Directions of Technical Change', *Research Policy*, 11(3): 147–62.

—— (1988) 'The Nature of the Innovative Process', in G. Dosi, C. Freeman, R. Nelson, G. Silverberg, and L. Soete (eds.), *Technical Change and Economic Theory*, London: Pinter, 221–38.

—— and Orsenigo, L. (1988) 'Coordination and Transformation: An Overview of Structures, Behaviours and Change in Evolutionary Environments', in G. Dosi, C. Freeman, R. Nelson, G. Silverberg, and L. Soete (eds.), *Technical Change and Economic Theory*, London: Pinter, 13–37.

—— Freeman, C., and Fabiani, S. (1994) 'The Process of Economic Development: Introducing Some Stylized Facts and Theories of Technologies, Firms and Institutions', *Industrial and Corporate Change*, 3(1): 1–45.

—— Marsili, O., Orsenigo, L., and Salvatore, R. (1993) 'Learning, Market Selection and the Evolution of Industrial Structures', paper presented at the EUNETIC Conference, Evolutionary Economics of Technical Change: Assessment of Results and New Frontiers, European Parliament, Strasbourg, 6–8 Oct.

—— Teece, D. J., and Winter, S. (1992) 'Toward a Theory of Corporate Coherence: Preliminary Remarks', in G. Dosi, R. Giannetti, and P. A. Toninelli (eds.), *Technology and Enterprise in a Historical Perspective*, Oxford: Clarendon Press, 184–211.

Douglas, M., and Isherwood, B. (1979) *The World of Goods: Towards an Anthropology of Consumption*, Harmondsworth: Penguin.

Downing, J., Mohammadi, A., and Sreberny-Mohammadi, A. (eds.) (1990) *Questioning the Media: A Critical Introduction*, London: Sage Publications.

Drake, W. J., and Nicolaidis, K. (1992) 'Ideas, Interests, and Institutionalization: "Trade in Services" and the Uruguay Round', *International Organization*, 46(1): 37–100.

Drucker, P. F. (1969) *The Age of Discontinuity: Guidelines to Our Changing Society*, New York: Harper & Row.

Dumas, A., and Mintzberg, H. (1989) 'Managing Design, Designing Management', *Design Management Journal*, Fall, 37–43.

Dunning, J. H. (1992) *Multinational Enterprises and the Global Economy*, Wokingham: Addison-Wesley.

Dutton, W. H. (1992) 'The Ecology of Games Shaping Telecommunications Policy', *Communication Theory*, 2(4): 303–28.

—— and Kraemer, K. (1985) *Modelling as Negotiating: The Role of Computer Models in the Policy Process*, Norwood, NJ: Ablex Publishing.

—— Blumler, J. G., and Kraemer, K. L. (eds.) (1987) *Wired Cities: Shaping the Future of Communications*, Boston: G. K. Hall & Co.

—— Taylor, J., Bellamy, C., Raab, C., and Peltu, M. (1994) 'Electronic Service Delivery: Themes and Issues in the Public Sector, A Forum Discussion', PICT Policy Research Paper No. 28, Programme on Information and Communication Technologies, Brunel University, 4 Mar.

Edge, D. (1988) *The Social Shaping of Technology*, Edinburgh PICT Working Papers No. 1, University of Edinburgh, Edinburgh.

Edquist, C., and Johnson, B. (1995) 'Institutions and Innovations: Conceptual Discussion', paper presented at the Systems of Innovation Research Network Workshop, Canary Island, 20–2 Jan.

Egan, B. L. (1994) 'Building Value through Telecommunications: Regulatory Roadblocks on the Information Superhighway', *Telecommunications Policy*, 18(8): 573–87.

Eisenstein, E. (1979) *The Printing Press as an Agent of Social Change*, 2 vols., Cambridge: Cambridge University Press.

EKOS Research Associates (1993) *Privacy Revealed: The Canadian Privacy Survey*, Ottawa: EKOS Research Associates.

European Commission (1987) 'Towards a Dynamic European Economy: Green Paper on the Development of the Common Market for Telecommunications Services and Equipment', European Commission, COM(87) 290 final, Brussels.

—— (1990) 'Commission Green Paper on the Development of European Standardization: Action for Faster Technological Integration in Europe', European Commission, COM(90) 456 final, Brussels.

—— (1994) 'EU Electronic Information Supply Industry Statistics in Perspective', Information Market Observatory, IMO Working Paper 94/5, Luxembourg, Nov.

Ewen, S. (1984) *All Consuming Images*, New York: Basic Books.

—— and Ewen, E. (1982) *Channels of Desire*, New York: McGraw-Hill.

Farrell, J., and Saloner, G. (1988) 'Coordination through Committees and Markets', *Rand Journal of Economics*, 19(2): 235–52.

Featherstone, M. (1991) *Consumer Culture and Postmodernism*, London: Sage.

Federal Communications Commission (1994) 'Report and Order and Further Notice of Proposed Rulemaking in the Matter of Rules and Policies regarding Calling Number Identification Service—Caller ID (FCC 94–59)', Washington, DC: Federal Communications Commission.

Ferguson, M. (1986) 'The Challenge of Neo-technological Determinism for Communication Systems, Industry and Culture', in M. Ferguson (ed.), *New*

Communication Technologies and the Public Interest: Comparative Perspectives on Policy and Research, London: Sage, 52–70.

Finkelstein, L. S. (1991) 'What is International Governance?' Paper prepared for the Annual Meeting of the International Studies Association, Vancouver, BC, 21 Mar.

Finnie, G. (1993) 'Top 100', *Communications Week International*, 14 Nov., 20–7.

Fischer, C. S. (1992) *America Calling: A Social History of the Telephone to 1940*, Berkeley and Los Angeles: University of California Press.

Flaherty, D. H. (1985) *Protecting Privacy in Two-Way Electronic Services*, White Plains, NY: Knowledge Industry Publications.

Fleck, J. (1994) 'Continuous Evolution—Corporate Configurations of Information Technology', in R. Mansell (ed.), *The Management of Information and Communication Technologies: Emerging Patterns of Control*, London: Aslib, 178–206.

Foray, D. (1994) 'Production and Distribution of Knowledge in the New Systems of Innovation: The Role of Intellectual Property Rights', *OECD STI Review*, 14: 119–52.

Forester, T. (ed.) (1989) *Computers in the Human Context: Information, Productivity and People*, Oxford: Blackwell.

Forty, A. (1986) *Objects of Desire: Design and Society 1750–1980*, London: Routledge & Kegan Paul, repr. 1972.

Foss, N. J. (1993) 'Theories of the Firm: Contractual and Competence Perspectives', *Journal of Evolutionary Economics*, 3: 127–44.

Foucault, M. (1977) *Discipline and Punish: The Birth of the Prison*, New York: Vintage.

Franz, C. R., and Robey, D. (1984) 'An Investigation of User-led Systems Design: Rational and Political Perspectives', *Communications of the ACM*, 27(12): 1202–9.

Freeman, C. (1982) *The Economics of Industrial Innovation*, London: Pinter.

—— (1991) 'Networks of Innovators: A Synthesis of Research Issues', *Research Policy*, 20(5): 499–514.

—— (1992a) 'The Human Use of Human Beings and Technical Change', in C. Freeman (1992) *The Economics of Hope: Essays on Technical Change, Economic Growth and the Environment*, London: Pinter, 175–89.

—— (1992b) 'Technology, Progress and the Quality of Life', in C. Freeman (1992) *The Economics of Hope: Essays on Technical Change, Economic Growth and the Environment*, London: Pinter, 212–30.

—— (1992c) 'Formal Scientific and Technical Institutions in the National System of Innovation', in B.-A. Lundvall (ed.), *National Systems of Innovation: Towards a Theory of Innovation and Interactive Learning*, London: Pinter, 169–87.

—— (1994a) 'The Diffusion of Information and Communication Technology in the World Economy in the 1990s', in R. Mansell (ed.), *The Management of Information and Communication Technologies: Emerging Patterns of Control*, London: Aslib, 8–41.

—— (1994b) 'The Economics of Technical Change, Critical Survey', *Cambridge Journal of Economics*, 18: 463–514.

—— (1994c) 'Marching to the Sound of a Different Drum', paper presented at

238   *Bibliography*

the EUNETIC Conference, Evolutionary Economics of Technical Change: Assessment of Results and New Frontiers, European Parliament, Strasbourg, 6–8 Oct.

Freeman, C., and Perez, C. (1988) 'Structural Crises of Adjustment, Business Cycles and Investment Behaviour', in G. Dosi, C. Freeman, R. Nelson, G. Silverberg, and L. Soete (eds.), *Technical Change and Economic Theory*, London: Pinter, 38–66.

—— and Soete, L. (1994) *Work for all or Mass Unemployment? Computerised Technical Change into the 21st Century*, London: Pinter.

Friedman, A. L. (1993) 'The Information Technology Field: An Historical Analysis', in P. Quintas (ed.), *Social Dimensions of Systems Engineering*, Hemel Hempstead: Ellis Horwood, 18–33.

—— with Cornford, D. S. (1989) *Computer Systems Development: History, Organisation and Implementation*, Chichester: Wiley.

Frith, S. (1983) 'The Pleasures of the Hearth', in *Formations of Pleasure*, London: Routledge & Kegan Paul.

Gandy (Jr.), O. H. (1993) *The Panoptic Sort: A Political Economy of Personal Information*, Boulder, Colo: Westview.

Gardner, C. B. (1988) 'Access Information: Public Lies and Private Peril', *Social Problems*, 35(4): 384–97.

Garnham, N. (1994) 'Whatever Happened to the Information Society?' in R. Mansell (ed.), *The Management of Information and Communication Technologies: Emerging Patterns of Control*, London: Aslib, 42–51.

Gell, A. (1986) 'Newcomers to the World of Goods: Consumption among the Muria Gonds', in A. Appadurai (ed.), *The Social Life of Things: Commodities in Cultural Perspective*, Cambridge: Cambridge University Press, 110–38.

General Agreement on Tariffs and Trade (GATT) (1986) 'Ministerial Declaration on the Uruguay Round', GATT Press Release No. 1396, Geneva, 25 Sept.

—— (1989) 'Trade in Telecommunication Services', Note by the Secretariat, Group of Negotiations on Services, MTN.GNS/W/52, Geneva, May.

—— (1994) *The Results of the Uruguay Round of Multilateral Trade Negotiations: The Legal Texts*, Geneva: GATT Secretariat.

Genschel, P., and Werle, R. (1992) 'From Hierarchical Coordination to International Standardization: Historical and Modal Changes in the Coordination of Telecommunications', Max-Planck-Institut fur Gesellschaftsforschung, Working Paper, Cologne.

Gerbner, G. (1986) 'Living with Television: The Dynamics of the Cultural Process', in J. Bryant and D. Zillman (eds.), *Perspectives on Media Effects*, Hillside, NJ: Lawrence Erlbaum, 17–40.

Geroski, P. A. (1990) 'Procurement Policy as a Tool of Industrial Policy', *International Review of Industrial Economics*, 4(2): 182–98.

Gershuny, J., and Miles, I. (1983) *The New Service Economy*, London: Pinter.

Gibbons, M., Limoges, C., Nowotny, H., Schwartzman, S., Scott, P., and Trow, M. (1994) *The New Production of Knowledge: The Dynamics of Science and Research in Contemporary Societies*, London: Sage.

Gibson, W. (1984) *Neuromancer*, New York: Ace Books.

—— (1987) *Count Zero*, New York: Ace Books.

—— (1988) *Mona Lisa Overdrive*, New York: Bantam Books.

Giddens, A. (1979) *Central Problems in Social Theory*, Berkeley and Los Angeles: University of California Press.

—— (1981) *A Contemporary Critique of Historical Materialism*, i, Berkeley and Los Angeles: University of California Press.

—— (1984) *The Constitution of Society: Outline of the Theory of Structuration*, London: Polity Press.

—— (1987) *The Nation State and Violence: Volume Two of a Contemporary Critique of Historical Materialism*, Berkeley and Los Angeles: University of California Press.

—— (1989) 'A Reply to my Critics', in D. Held and J. Thompson (eds.), *Social Theory of Modern Societies: Anthony Giddens and his Critics*, Cambridge: Cambridge University Press, 249–301.

—— (1991) *Modernity and Self-Identity: Self and Society in the Late Modern Age*, Stanford, Calif.: Stanford University Press.

Giedion, S. (1941) *Space, Time and Architecture: The Growth of a New Tradition*. Cambridge, Mass./London: Harvard University Press/Oxford University Press, repub. 1967.

Gilpin, R. (1987) *The Political Economy of International Relations*, Princeton, NJ: Princeton University Press.

Glaser, B., and Strauss, A. (1968) *The Discovery of Grounded Theory*, London: Weidenfeld & Nicholson.

Glegg, G. L. (1968) *The Design of Design*, Cambridge: Cambridge University Press.

Gnoffo, A. (1990) 'They've Got Your Number', *Philadelphia Inquirer*, 4 Feb.

Goffman, E. (1963) *Behavior in Public Places: Notes on the Social Organization of Gatherings*, New York: Free Press.

—— (1971) *Relations in Public: Microstudies of the Public Order*, New York: Basic Books.

Goldberg, H. (1985) 'One-Hundred and Twenty Years of International Communications', *Federal Communications Law Journal*, 37: 131–54.

Golding, P., and Murdock, G. (1978) 'Theories of Communication and Theories of Society', *Communication Research*, 5(3): 339–56.

Gore, A. (1994). Text of speech delivered by the Vice-President of the United States of America to the Meeting of the International Telecommunication Union, Buenos Aires, Argentina, 21 Mar.

—— (1993) Speech by Vice-President Albert Gore, Academy of Television, Arts and Science, Los Angeles, Nov.

Gottdiener, M. (1986) 'Recapturing the Center: A Semiotic Analysis of Shopping Malls', in M. Gottdiener and A. P. Lagopoulos (eds.), *The City and the Sign: An Introduction to Urban Semiotics*, New York: Columbia University Press, 288–302.

Gough, I. (1994) 'Economic Institutions and the Satisfaction of Human Needs', *Journal of Economic Issues*, 28(1): 25–66.

Gregory, D. (1981) 'Human Agency and Human Geography', *Transactions of the Institute of British Geographers*, 6(1): 1–18.

—— (1989) 'Presences and Absences: Time-Space Relations and Structuration Theory', in D. Held and J. B. Thompson (eds.), *Social Theory of Modern Societies: Anthony Giddens and his Critics*, Cambridge: Cambridge University Press, 185–214.

Gregory, D. and Urry, J. (eds.) (1985) *Social Relations and Spatial Structures*, London: Macmillan.

Group of Prominent Persons (1994) 'Europe and the Global Information Society: Recommendations to the European Council', Industrial Group of Prominent Persons, Brussels, 26 May.

Group of Seven (1991) 'Economic Declaration of the Group of Seven, Article 11', Economic Summit, London, 17 July.

Gumpert, G. (1992) 'The Ambiguity of Perception', in D. Shimkin, H. Stolerman, and H. O'Conor (eds.), *State of the Art: Issues in Contemporary Mass Communication*, New York: St Martin's Press, 378–87.

Guzzini, S. (1993) 'Structural Power: The Limits of Neorealist Power Analysis', *International Organization*, 47(3): 443–78.

Haas, E. B. (1964) *Beyond the Nation State*, Stanford, Calif: Stanford University Press.

—— (1990) *When Knowledge is Power: Three Models of Change in International Organizations*, Berkeley and Los Angeles: University of California Press.

Haas, P. M. (1992) 'Introduction: Epistemic Communities and International Policy Coordination', *International Organization*, 46(1): 1–36.

Haddon, L. (1988) 'The Home Computer: The Making of a Consumer Electronic', *Science as Culture*, 2: 7–51.

—— (1992) 'Explaining ICT Consumption: The Case of the Home Computer', in R. Silverstone and E. Hirsch (eds.), *Consuming Technologies: Media and Information in Domestic Spaces*, London: Routledge, 82–96.

—— and Silverstone, R. (1993) 'Teleworking in the 1990s: A View from the Home', SPRU CICT Report No. 10, Brighton: Science Policy Research Unit, University of Sussex.

—— and Silverstone, R. (1994) 'The Careers of Information and Communication Technologies in the Home', in K. Bjerg and K. Borreby (eds.), *HOIT94 Proceedings: Home Oriented Informatics Telematics and Automation*, Copenhagen, 275–84.

—— and Silverstone, R. (1995) *Information and Communication Technologies and the Young Elderly*, SPRU CICT Report No 13, Brighton: Science Policy Research Unit, University of Sussex.

Hales, M. (1993) 'User Participation in Design—What It Can Deliver, What It Can't and What This Means for Management', in P. Quintas (ed.), *Social Dimensions of System Engineering: People, Processes, Policies and Software Development*, Hemel Hempstead: Ellis Horwood, 215–35.

Ham, C., and Hill, M. (1984) *The Policy Process in the Modern Capitalist State*, 2nd edn., London: Harvester Wheatsheaf.

Hammer, M., and Champy, J. (1993) *Re-Engineering the Corporation—A Manifesto for Business Revolution*, London: Nicholas Brearly Publishing.

Harasim, L. M. (ed.) (1993) *Global Networks: Computers and International Communication*, Cambridge, Mass: MIT Press.

Hart, J., Reed, R., and Bar, F. (1992) 'The Future of Networking in the US: The Building of the Internet: Implications for the Future of Broadband Networks', Report on the Future of Networking in the United States prepared for the Commission of the European Communities, Brussels.

Harvey, D. (1973) *Social Justice and the City*, London: Edward Arnold.

—— (1985) *The Urbanisation of Capital*, Oxford: Blackwell.

—— (1989) *The Condition of Postmodernity: An Enquiry into the Origins of Cultural Change*, Oxford: Blackwell.

Hawkins, R. W. (1992) 'The Doctrine of Regionalism: A New Dimension for International Standardization in Telecommunication', *Telecommunications Policy*, 16(4): 339–53.

—— (1993*a*) 'Public Standards and Private Networks: Some Implications of the Mobility Imperative', European Network for Communication and Information Perspectives (ENCIP), Working Paper, Mar., Montpellier.

—— (1993*b*) 'Changing Expectations: Voluntary Standards and the Regulation of European Telecommunication', *Communications & Strategies*, 11: 53–85.

—— (1994*a*) 'Regional Technical Infrastructures: The European Position in the Evolving Geo-Politics of Telecommunication', European Network for Communication and Information Perspectives (ENCIP) Working Paper, July, Montpellier.

—— (1994*b*) 'Facilitator's Report, International Governance of Technology', Science Policy Research Unit, University of Sussex, Brighton, mimeo.

—— (1995*a*) 'Standards-Making as Technological Diplomacy: Assessing Objectives and Methodologies in Standards Institutions', in R. Hawkins, R. Mansell, and J. Skea (eds.), *Standards, Innovation and Competitiveness: The Politics and Economics of Standards in Natural and Technical Environments*, Cheltenham: Edward Elgar, 147–58.

—— (1995*b*) 'The User Role in the Development of Technical Standards for Telecommunication', *Technology Analysis and Strategic Management*, 7(1): 21–40.

—— Mansell, R., and Skea, J. (eds.) (1995) *Standards, Innovation and Competitiveness: The Politics and Economics of Standards in Natural and Technical Environments*, Cheltenham: Edward Elgar.

Hayek, F. (1945) 'The Use of Knowledge in Society', *American Economic Review*, 35: 519–30.

Hemenway, D. (1975) *Industrywide Voluntary Product Standards*, Cambridge, Mass.: Ballinger.

—— (1980) *Performance vs Design Standards*, Washington, DC: Dept. of Commerce, National Bureau of Standards.

Hills, J. (1986) *Deregulating Telecoms: Competition and Control in the United States, Japan and Britain*, Westport, Conn.: Quorum Books.

Hirscheim, R., and Klein, H. K. (1989) 'Four Paradigms of Information Systems Development', *Communications of the ACM*, 32(10): 1199–216.

Hirshleiffer, J. (1973) 'Where Are We in The Theory of Information?', *American Economic Review*, 63(3): 31–9.

—— and Riley, J. G. (1979) 'The Analytics of Uncertainty and Information—An Expository Survey', *Journal of Economic Literature*, 17: 1375–417.

Hodgson, G. M. (1989) 'Institutional Economic Theory: The Old Versus the New', *Review of Political Economy* 1(3): 249–69.

—— Samuels, W., and Tool, M. (eds.) (1994) *Institutional and Evolutionary Economics*, Cheltenham: Edward Elgar.

Hoffman, D. L., and Novak, T. P. (1994) 'Wanted: Net Census', *Wired*, Nov.: 93–4.

Hogwood, B. W., and Gunn, L. A. (1984) *Policy Analysis for the Real World*, Oxford: Oxford University Press.

Hootman, J. (1972) 'The Computer Network as a Marketplace', *Datamation*, 43–6.

Horwitz, R. B. (1993) 'South African Telecommunications: History and Prospects', La Jolla, Calif., University of California, unpub. mimeo.

House of Commons Trade and Industry Committee (1994) *Optical Fibre Networks*, London: HMSO.

Huber, P. W. (1993) 'Telecommunications Regulation: The Beginning of the End', *Issues in Science and Technology*, Fall: 50–8.

Humphrey, W. S. (1989) *Managing the Software Process*, Reading, Mass: Addison-Wesley.

Huth, V., and Gould, S. B. (1993) 'The National Information Infrastructure: The Federal Role', Congressional Research Service Issue Brief, Washington, DC: Library of Congress, Congressional Research Service.

Imai, K., and Itami, H. (1984) 'Interpenetration of Organization and Market', *International Journal of Industrial Organization*, 2: 285–310.

Innis, H. A. (1950) *Empire and Communication*, Toronto: Oxford University Press.

—— (1951) *The Bias of Communication*, Toronto: University of Toronto Press.

International Telecommunication Union (1988) 'Final Acts of the World Administrative Telegraph and Telephone Conference, International Tele-communication Regulations', Art. 9 and Opinion PL/A International Tele-communication Union, Geneva, Dec.

—— (1989) 'The Changing Telecommunication Environment: Policy Considera-tions for Members of the ITU', Advisory Group on Telecommunication Policy, International Telecommunication Union, Geneva, Feb.

—— (1993) 'Tomorrow's ITU: The Challenges of Change', Report of the High Level Committee to Review the Structure and Function of the ITU, International Telecommunication Union, Geneva.

—— (1994) 'World Telecommunication Development Report 1994: World Tele-communication Indicators', International Telecommunication Union, Geneva, Mar.

Jackson, M. (1994) 'Problems, Methods and Specialisation', *Software Engineering Journal*, 9(6): 249–55.

James, C. L., and Morrill, D. E. (1983) 'The Real Ada, Countess of Lovelace', *ACM SIGSOFT Software Engineering Notes*, 8(1): 30–1.

Jasanoff, S. (1986) *Risk Management and Political Culture*, New York: Sage.

—— (1987) 'Contested Boundaries in Policy-Relevant Science', *Social Studies of Science*, 17: 195–230.

Jhally, S. (1987) *The Codes of Advertising: Fetishism and the Political Economy of Meaning in the Consumer Society*, London: Routledge.

Johanson, J., and Mattsson, L.-G. (1994) 'The Markets-as-Networks Tradition in Sweden', in G. Laurent, G. L. Lilien, and B. Pras (eds.), *Research Traditions in Marketing*, Amsterdam: Kluwer Academic Publishers, 321–42.

Johnson, B. (1992) 'Institutional Learning', in B.-A. Lundvall (ed.), *National Systems of Innovation: Towards a Theory of Innovation and Interactive Learning*, London: Pinter, 23–44.

Jones, C. (1993) *Software Productivity and Quality Today: A Worldwide Perspective*, Carlsbad, Calif.: IS Management Group.

Jonscher, C. (1983) 'Information Resources and Economic Productivity', *Information Economics and Policy*, 1: 13–35.

Jouet, J. (1993) 'A Telematic Community: The Axiens', in J. Jouet, P. Flichy, and P. Beaud (eds.), *European Telematics: The Emerging Economy of Words*, Amsterdam: North-Holland, 181–202.

Kaplan, D. (1990) *The Crossed Line: The South African Telecommunications Industry in Transition*, Johannesburg: Witwatersrand University Press.

Kapor, M. (1993) 'Where is the Digital Highway Really Heading? The Case for a Jeffersonian Information Policy', *Wired*, July–Aug., 53–9, 94.

Katz, M. L., and Shapiro, C. (1986) 'Technology Adoption in the Presence of Network Externalities', *Journal of Political Economy*, 94: 882–941.

Keen, B. (1987) ' "Play It Again Sony": The Double Life of Home Video Technology', *Science as Culture*, 1: 7–42.

Kidder, T. (1981) *The Soul of a New Machine*, Harmondsworth: Penguin.

Kimberley, P. (1991) *Electronic Data Interchange*, New York: McGraw-Hill.

Kimble, C., and McLoughlin, K. (1994) 'Changes to the Organisation and the Work of Managers Following the Introduction of an Integrated Information System', in R. Mansell (ed.), *The Management of Information and Communication Technologies: Emerging Patterns of Control*, London: Aslib, 158–77.

Kindleberger, C. P. (1983) 'Standards as Public, Collective, and Private Goods', *Kyklos*, 26: 377–96.

Kling, R., and Iacono, S. (1984) 'The Control of Information System Developments after Implementation', *Communications of the ACM*, 27(12): 1218–26.

Knorr-Cetina, K., and Cicourel, A. (eds.) (1981) *Advances in Social Theory: Toward an Integration of Micro and Macro-Sociologies*, London: Routledge & Kegan Paul.

Kopytoff, I. (1986) 'The Cultural Biography of Things: Commoditization as Process', in A. Appadurai (ed.), *The Social Life of Things: Commodities in Cultural Perspective*, Cambridge: Cambridge University Press, 64–91.

Kozmetsky, G., and Kircher, P. (1956) *Electronic Computers and Management Control*, New York: McGraw-Hill.

Krasner, S. D. (1991) 'Global Communications and National Power: Life on the Pareto Frontier', *World Politics*, 43(Apr.), 336–66.

Kratchowil, F., and Ruggie, J. G. (1986) 'International Organization: A State of the Art on an Art of the State', *International Organization*, 40(4): 753–75.

Kraut, R. E. (1994) 'Prospects for Video', in S. J. Emmott and D. S. Travis (eds.), *POTS to PANS Symposium*, Hintlesham, Suffolk, n.p.

Lacey, K. (1994) 'From Plauderei to Propaganda: On Women's Radio in Germany 1924–35', *Media, Culture and Society*, 4(16): 589–608.

Lamberton, D. (1984) 'The Emergence of Information Economics', in M. Jussawalla and H. Ebenfield (eds.), *Communication and Information Economics*, Amsterdam: North-Holland, 7–22.

—— Macdonald, S. and Mandeville, T. (1986) 'Information and Technological Change—A Research Program in Retrospect', in P. Hall (ed.), *Technology, Innovation and Economic Policy*, Oxford: Philip Allan, 231–43.

Lammers, S. (1986) *Programmers at Work*, Seattle: Microsoft Press.

Lane, R. E. (1966) 'The Decline of Politics and Ideology in a Knowledgeable Society', *American Sociological Review*, 31, 649–62.

Lane, R. E. (1990) 'Concrete Theory: An Emerging Political Method', *American Political Science Review*, 84(3): 927–41.

Langlois, R. N. (1989) 'What Was Wrong with the Old Institutional Economics (and What Is still Wrong with the New)?', *Review of Political Economy*, 1(3): 270–98.

Larson, E. (1992) *The Naked Consumer*, New York: Henry Holt.

Lasch, C. (1991) *The True and Only Heaven*, New York/London: W. W. Norton & Co.

Law, J. (1987) 'Technology and Heterogeneous Engineering: The Case of the Portuguese Expansion', in W. E. Bijker, T. Hughes, and T. Pinch (eds.), *The Social Construction of Technical Systems*, Cambridge, Mass.: MIT Press, 111–34.

—— and Bijker, W. E. (1992) 'Postscript: Technology, Stability, and Social Theory', in W. E. Bijker and J. Law (eds.), *Shaping Technology/Building Society: Studies in Sociotechnical Change*, Cambridge, Mass.: MIT Press, 290–308.

Layder, D. (1981) *Structure, Interaction, and Social Theory*, London: Routledge & Kegan Paul.

Leathers, C. G. (1989) 'New and Old Institutionalists on Legal Rules: Hayek and Commons', *Review of Political Economy* 1(3): 361–80.

Lefebvre, H. (1991) *The Production of Space*, trans. by D. Nicholson-Smith, Oxford: Blackwell, originally pub. 1974.

Lehman, B. A. (1994) 'Green Paper—Intellectual Property and the National Information Infrastructure, A Preliminary Draft of the Report of the Working Group on Intellectual Property Rights', Information Infrastructure Task Force, Washington, DC, July.

Leiss, W. (1976) *The Limits to Satisfaction*, London: Marion Boyars.

—— Kline, S. and Jhaly, S. (1990) *Social Communication in Advertising: Persons, Products and Images of Well-Being*, London: Routledge.

Lemoine, P. (1988) 'The Demise of Classical Rationality', in J. Thackara (ed.), *Design after Modernism: Beyond the Object*, London: Thames & Hudson, 187–96.

Leonard-Barton, D. (1988) 'Implementation as Mutual Adaptation of Technology and Organization', *Research Policy*, 17(5): 251–67.

—— (1992) 'Core Capabilities and Core Rigidities: A Paradox in Managing New Product Development', *Strategic Management Journal*, 13: 111–25.

Levy, C. J. (1991) 'Who's Being Helped by Help Lines?', *New York Times*, 14 May.

Lewyn, M. (1992) 'Phone Sleuths are Cutting off Hackers', *Business Week*, 13 July.

Lientz, B. P., and Swanson, E. B. (1980) *Software Maintenance Management*, Reading, Mass: Addison-Wesley.

Lievrouw, L. A., and Finn, T. A. (1990) 'Identifying the Common Dimensions of Communication: The Communication Systems Model', in B. D. Ruben and L. A. Lievrouw (eds.), *Information and Behavior*, iii, Brunswick, NJ: Transaction Books, 37–65.

Liff, S., and Scarbrough, H. (1994) 'Creating a Knowledge Database— Operationalising the Vision or Compromising the Concept?', in R. Mansell (ed.) *The Management of Information and Communication Technologies: Emerging Patterns of Control*, London: Aslib, 207–19.

Lipsey, R. (1994) 'Sustainable Growth, Innovation, Competitiveness and

Foreign Trade', background paper for the Trinational Institute on Innovation, Competitiveness and Sustainability, Whistler, BC, 14 Aug.

Lloyd, C. (1986) *Explanation in Social History*, Oxford: Blackwell.

Louis Harris and Associates (1993) *Health Care Information Privacy: A Survey of the Public and their Leaders Conducted for Equifax Inc.*, Atlanta: Equifax.

Low, J., and Woolgar, S. (1993) 'Managing the Socio-Technical Divide: Some Aspects of the Discursive Structure of Information Systems', in P. Quintas (ed.), *Social Systems of Software Engineering: People, Processes, Policies and Software Development*, Hemel Hempstead: Ellis Horwood, 34–59.

Luard, E. (1977) *International Agencies: The Emerging Framework of Interdependence*. London: Macmillan (Royal Institute of International Affairs).

—— (1990) *The Globalization of Politics: The Changed Focus of Political Action in the Modern World*, London: Macmillan.

Lundvall, B.-A. (1988) 'Innovation as an Interactive Process: From User-Producer Interaction to National Systems of Innovation', in G. Dosi, C. Freeman, R. Nelson, G. Silverberg, and L. Soete (eds.), *Technical Change and Economic Theory*, London: Pinter, 349–69.

—— (ed.) (1992*a*) *National Systems of Innovation: Towards a Theory of Innovation and Interactive Learning*, London: Pinter.

—— (1992*b*) 'Introduction', in B.-A. Lundvall (ed.), *National Systems of Innovation: Towards a Theory of Innovation and Interactive Learning*, London: Pinter, 1–19.

—— (1992*c*) 'User-Producer Relationships, National Systems of Innovation and Internationalisation', in B.-A. Lundvall (ed.), *National Systems of Innovation: Towards a Theory of Innovation and Interactive Learning*, London: Pinter, 45–67.

Lyon, D. (1986) 'From "Post-Industrialism" to "Information Society": A New Social Transformation?', *Sociology*, 20: 577–88.

—— (1994) *The Electronic Eye: The Rise of Surveillance Society*, London: Sage.

McCracken, G. (1988) *Culture and Consumption*, Bloomington, Ind.: Indiana University.

McDowell, S. D. (1994) 'International Services Liberalisation and Indian Telecommunications Policy', in E. A. Comor (ed.), *The Global Political Economy of Communication: Hegemony, Telecommunication and the Information Economy*, New York: St Martin's Press, 103–24.

Machlup, F. (1962) *The Production and Distribution of Knowledge*, Princeton, NJ: Princeton University Press.

Mackay, H. (1995) 'Patterns of Ownership of IT Devices in the Home', in N. Heap, R. Thomas, G. Einon, R. Mason, and H. Mackay (eds.), *Information Technology and Society*, London: Sage, 311–40.

McKee, J. R. (1984) 'Maintenance is a Function of Design', in *Proceeding of the 1984 AFIPS National Computer Conference*, 187–93.

McKelvey, M. (1991) 'How do National Systems of Innovation Differ? A Critical Analysis of Porter, Freeman, Lundvall and Nelson', in G. Hodgson and E. Screpanti (eds.), *Rethinking Economics: Markets, Technology and Economic Evolution*, Cheltenham: Edward Elgar, 117–37.

MacKenzie, D. (1993) 'Negotiating Arithmetic, Constructing Proof: The Sociology of Mathematics and Information Technology', *Social Studies of Science*, 23: 37–65.

Mackenzie, D. and Wajcman, J. (eds.) (1985) *The Social Shaping of Technology*, Milton Keynes: Open University Press.

McLuhan, M. (1951*a*) 'Introduction', in H. A. Innis (1951), pp. vii–xviii.

—— (1951*b*) *The Mechanical Bride*, New York: Vanguard Press.

—— (1964) *Understanding Media: The Extensions of Man*, New York: McGraw-Hill.

McManus, T. (1990) *Telephone Transaction-Generated Information: Rights and Restrictions*, Cambridge, Mass.: Harvard University Center for Information Policy Research.

Macro, A., and Buxton, J. N. (1987) *The Craft of Software Engineering*, Wokingham: Addison-Wesley.

Mai, M. (1988) 'Soziologische Fragen der Technischen Normung', *Sozialwissenschaften und Berufspraxis*, 11(2): 115–27.

Majone, G. (1984) 'Science and Trans-Science in Standard Setting', *Science, Technology and Human Values*, 9(1): 15–22.

—— (1989) *Evidence, Argument and Persuasion in the Policy Process*, New Haven: Yale University Press.

Malone, T. W., Benjamin, R. I., and Yates, J. (1987) 'Electronic Markets and Electronic Hierarchies', *Communications of the ACM*, 30(6): 484–97.

Mansell, R. (1990) 'Rethinking the Telecommunication Infrastructure: The New "Black Box" ', *Research Policy*, 19(6): 501–15.

—— (1993*a*) *The New Telecommunications: A Political Economy of Network Evolution*, London: Sage.

—— (1993*b*) 'Trans-European Telecommunication: User Perspectives on the Prospects and Problems', European Network for Communication and Information Perspectives (ENCIP), Working Paper No. 4, Montpellier, Feb.

Emerging Patterns of Control, London: Aslib.

—— (1995) 'Innovation in Telecommunication: Bridging the Supplier-User Interface', in M. Dodgson and R. Rothwell (eds.), *Handbook of Industrial Innovation*, Cheltenham: Edward Elgar, 232–42.

—— and Credé, A. (1995) 'Telecommunication Competition and "Commodity" Supply', European Network for Communication and Information Perspectives (ENCIP), Working Paper, Montpellier.

—— and Hawkins, R. (1992) 'Old Roads and New Signposts: Trade Policy Objectives in Telecommunication Standards', in F. Klaver and P. Slaa (eds.), *Telecommunication: New Signposts to Old Roads*, Amsterdam: IOS Press, 45–54.

—— and Jenkins, M. (1992*a*) 'Electronic Trading Networks and Interactivity: The Route Toward Competitive Advantage?', *Communications & Strategies*, 6: 63–85.

—— —— (1992*b*) 'Networks and Policy: Interfaces, Theories and Research', *Communications & Strategies*, 5: 31–50.

—— —— (1993) 'Coherence and Diversity of Systems of Innovation: Electronic Trading Systems—The Emergence of New Productive Systems', Report prepared for the Commission of the European Communities FAST project, LATAPSES, Sophia Antipolis, published by the European Commission, Brussels, Feb.

—— and Tang, P. (1994) 'Electronic Information Services: Competitiveness in the United Kingdom', Report prepared for the Department of Trade and

Industry, Textiles and Retailing Division, by Science Policy Research Unit, University of Sussex, Brighton, July.

—— Hawkins, R., and Jenkins, M. (1991) 'Technical and Policy Prerequisites for Global Networking Efficiency', Report prepared for the UNCTAD Special Programme on Trade Facilitation, Geneva, Science Policy Research Unit, University of Sussex, Brighton, Dec.

March, J. G. (1978) 'Bounded Rationality, Ambiguity, and the Engineering of Choice', *Bell Journal of Economics*, 9: 587–608.

—— and Olsen, J. P. (1989) *Rediscovering Institutions: The Organizational Basis of Politics*, New York: Free Press.

—— and Simon, H. A. (1967) *Organizations*, New York: John Wiley.

Marcuse, H. (1960) *Reason and Revolution: Hegel and Rise of Social Theory*, Boston: Beacon Press.

Markus, M. L., and Bjørn-Andersen, N. (1987) 'Power Over Users: Its Exercise by System Professionals', *Communications of the ACM*, 30(6): 498–504.

Marshal, E. (1989) 'NSF Opens High-Speed Computer Network', *Science*, 243(4887): 22–3.

Martinet, A. (1969) *Elements of General Linguistics*, London: Faber & Faber.

Marvin, C. (1988) *When Old Technologies Were New: Thinking About Communications in the Late Nineteenth Century*, Oxford: Oxford University Press.

Mattelart, A. (1991) *Advertising International: The Privatisation of Public Space*, London: Routledge.

Meessen, K. M. (1987) 'Intellectual Property Rights in International Trade', *Journal of World Trade Law*, 21(1): 67–74.

Melody, W. H. (1981) 'The Economics of Information as Resource and Product', in D. J. Wedemeyer (ed.), *PTC'81 Pacific Telecommunications Conference*, Honolulu: Pacific Telecommunications Council, C7-5-9.

—— (1987) 'Information: An Emerging Dimension of Institutional Analysis', *Journal of Economic Issues*, 21(3): 1313–39.

—— and Mansell, R. E. (1983) 'The Debate over Critical vs. Administrative Research: Circularity or Challenge', *Journal of Communication*, 33(3): 103–16.

Merton, R.K. (1968) *Social Theory and Social Structure*, New York: Free Press, (revd. edn.), (first pub. 1957).

Metcalfe, J. S. (1993) 'The Economic Foundations of Technology Policy: Equilibrium and Evolutionary Perspectives', final version, University of Manchester, mimeo, 92 pp.

—— and Boden, M. (1991) 'Innovation Strategy and the Epistemic Connection: An Essay on the Growth of Technological Knowledge', *Journal of Scientific & Industrial Research*, 50(Oct.), 707–17.

—— Birchenhall, C. R., and Kastrinos, N. (1994) 'Genetic Algorithms in Evolutionary Modelling: Abstract', for a paper for the EUNETIC Conference—Evolutionary Economics of Technical Change: Assessment of Results and New Frontiers European Parliament, Strasbourg, 6–8 Oct.

Meyrowitz, J. (1985) *No Sense of Place: The Impact of Electronic Media on Social Behaviour*, New York: Oxford University Press.

Miles, I., and Contributors (1990) *Mapping and Measuring the Information Economy: A Report Prepared for the ESRC Programme in PICT*, Library and Research Report 77, Boston Spa: British Library.

Miles, I., and Gershuny, J. (1986) 'The Social Economics of Information Technology', in M. Ferguson (ed.), *New Communication Technologies and the Public Interest*, London: Sage, 18–36.

—— and Matthews, M. (1992) 'Information Technology and the Information Economy', in K. Robins (ed.), *Understanding Information: Business, Technology and Geography*, London: Belhaven Press, 91–112.

—— Cawson, A., and Haddon, L. (1992) 'The Shape of Things to Consume', in R. Silverstone and E. Hirsch (eds.), *Consuming Technologies: Media and Information in Domestic Spaces*, London: Routledge, 67–81.

Miller, D. (1987) *Material Culture and Mass Consumption*, Oxford: Blackwell.

Miller, M. W. (1991) 'Lobbying Campaign, AT&T Directories Raise Fears about Use of Phone Records', *Wall Street Journal*, 13 Dec, B1.

Mitrany, D. (1966) *A Working Peace System*. Chicago: Quadrangle Books.

Molina, A. (1990) 'Transputers and Transputer-based Parallel Computers', *Research Policy*, 19(4): 309–35.

Monk, P. (1993) 'The Economic Significance of Infrastructural IT Systems', *Journal of Information Technology*, 8: 14–21.

Moore, N. (1991) 'Information Policy Research Priorities', *Policy Studies*, 15(1): 16–26.

Moores, S. (1988) ' "The Box on the Dresser": Memories of Early Radio and Everyday Life', *Media, Culture and Society*, 10(1): 10–23.

—— (1990) 'Texts, Readers and Contexts of Reading: Developments in the Study of Media Audiences', *Media, Culture and Society*, 12(1): 9–31.

Morley, D. (1992) 'Changing Paradigms in Audience Studies', in E. Seiter, H. Borchers, G. Kreutzner, and E.-M. Warth (eds.), *Remote Control: Television, Audiences and Cultural Power*, London: Routledge, 16–43.

—— and Silverstone, R. (1990) 'Domestic Communication: Technologies and Meanings', *Media, Culture and Society*, 12(1): 31–56.

Morris, L. (1990) *The Workings of the Household*, Cambridge: Polity Press.

Morse, C., Ashford, D. E., Bent, F. T., Friedland, W. H., Lewis, J. W., and Macklin, D. B. (1969) *Modernization by Design: Social Change in the Twentieth Century*, Ithaca, NY/London: Cornell University Press.

Motyka, C. (1992) 'US Participation in the Berne Convention and High Technology', *Copyright Law Symposium, New York*, 39: 107–39.

Mouzelis, N. (1980) 'Reductionism in Marxist Theory', *Telos*, 45(Fall), 173–85.

Mukherjee, R. (1994) 'The Communicative Processes of Regulation: A Study of State Commission Initiatives on Calling Line Identification Services', unpublished PhD Diss., Ohio State University.

—— and Samarajiva, R. (1993) 'The Customer Web: Transaction Generated Information and Telecommunication', *Media Information Australia*, 67: 51–61.

Mulgan, G. J. (1991) *Communication and Control: Networks and the New Economics of Communication*, London: Guildford Press.

Mumford, E. (1972) *Job Satisfaction: A Study of Computer Specialists*, London: Longman.

Murray, F., and Willmott, H. (1993) 'The Communications Problem in Information Systems Development: Towards a Relational Approach', in P. Quintas (ed.), *Social Dimensions of Systems Engineering: People, Processes, Policies and Software Development*, Hemel Hempstead: Ellis Horwood, 165–78.

National Research Council (1994) 'Realizing the Information Future', Computer Science and Telecommunications Board, Washington, DC, 25 May.

Neale, W. C. (1987) 'Institutions', *Journal of Economic Issues*, 21(3): 1177–206.

Nelson, R. (1988) 'Institutions Supporting Technical Change in the United States', in G. Dosi, C. Freeman, R. Nelson, G. Silverberg, and L. Soete (eds.), *Technical Change and Economic Theory*, London: Pinter, 312–29.

—— (ed.) (1993a) *National Innovation Systems: A Comparative Analysis*, Oxford: Oxford University Press.

—— (1993b) 'A Retrospective', in R. Nelson (ed.), *National Innovation Systems: A Comparative Analysis*, Oxford: Oxford University Press, 505–23.

—— (1994) 'The Co-evolution of Technology, Industrial Structure, and Supporting Institutions', *Industrial and Corporate Change*, 3(1): 47–63.

—— and Winter, S. (1977) 'In Search of a Useful Theory of Innovation', *Research Policy*, 6(1): 36–76.

—— —— (1982) *An Evolutionary Theory of Economic Change*, Cambridge: Belknap Press.

Newsweek (1995) 'United States Census 1993 Figures', *Newsweek*, 27 Feb.

Nimmer, D. (1992) 'Nation, Duration, Violation, Harmonization: International Copyright Proposal for the United States', *Law and Contemporary Problems*, 55(2): 211–39.

Niosi, J., Saviotti, P., Bellon, B., and Crow, M. (1993) 'National Systems of Innovation: In Search of a Workable Concept', *Technology in Society*, 15: 207–27.

Noam, E. M. (1987) 'The Public Telecommunications Network: A Concept in Transition', *Journal of Communication*, 37(1): 30–48.

—— (1991) 'Private Networks and Public Objectives', in *Universal Telephone Service: Ready for the 21st Century? Annual Review of the Institute for Information Studies*, Nashville and Queenstown, Md.: Northern Telecom and Aspen Institute, 1–27.

—— (1992) 'Network Tipping and the Tragedy of the Common Network: A Theory for the Formation and Breakdown of Public Telecommunication Systems', *Communications & Strategies*, special edn., 49–86.

—— (1994a) 'Beyond Liberalization II: The Impending Doom of Common Carriage', *Telecommunications Policy*, 18(6): 435–52.

—— (1994b) 'Beyond Liberalization: From the Network of Networks to the System of Systems', *Telecommunications Policy*, 18(4): 286–94.

—— (1994c) 'Beyond Liberalization III: Reforming Universal Service', *Telecommunications Policy*, 18(9): 687–704.

Noble, D. (1986) *Forces of Production: A Social History of Industrial Automation*, New York: Oxford University Press.

Nora, S., and Minc, A. (1980) *The Computerisation of Society*, Cambridge, Mass.: MIT Press, first pub. 1978 as *L'Informatisation de la société*.

Novek, E., Sinha, N., and Gandy (Jr.), O. H. (1990) 'The Value of your Name', *Media, Culture and Society*, 12: 524–43.

Oakley, B. (1990) 'Trends in the European IT Skills Scene', in *Proceedings of the Human Resource Development in Information Technology Conference*, Pergamon Infotech, London, 19–21 Feb., 1/1–1/4.

O'Connell, J. (1993) 'Metrology: The Creation of Universality by the Circulation of Particulars', *Social Studies of Science*, 23: 129–73.

O'Connor, R. (1984) 'The Use of Information Technology Standards in Public Procurement', Report to the Public Procurement Subcommittee, European Commission, Task Force for Information Technologies Telecommunication, Brussels.

—— (1992) 'Standards, Technical Regulations and Quality Assurance—What Will Change? What Implications for Community S&T Policy?', European Commission, DGXII, Luxembourg.

OECD (1989) 'Economic Arguments for Protecting Intellectual Property Rights Effectively', OECD, Paris.

—— (1994*a*) 'Communications Outlook 1994', OECD Working Party on Telecommunications and Information Services, DSTI/ICCP/TISP(94)5, OECD, Paris.

—— (1994*b*) *Information Technology Outlook*, Paris: OECD.

—— (1994*c*) 'OECD Statement on the Benefits of Telecommunications Infrastructure Competition', OECD Committee for Information, Computer and Communications Policy, Paris, 1 Apr.

Office of Technology Assessment (1986) 'Intellectual Property Rights in an Age of Electronic Information', Office of Technology Assessment, United States Congress, Washington, DC: Government Printing Office.

—— (1994) 'Electronic Enterprises: Looking to the Future', Office of Technology Assessment, United States Congress, Washington, DC, May.

Office of Telecommunications (1994) 'Households without a Telephone', Office of Telecommunications, London.

Ohio Bell Inc. (1991) 'Application for the Offering of Caller ID and Automatic Callback Services Before the Public Utilities Commission of Ohio, Case numbers 90-467-TP-ATA&90-471-TP-ATA', Columbus, Oh.: Ohio Public Utilities Commission.

Ohmae, K. (1990) *The Borderless World: Power and Strategy in the Interlinked Economy*, New York: Harper Business.

Olson, M. (1971) *The Logic of Collective Action: Public Goods and the Theory of Groups*, Cambridge, Mass.: Harvard University Press.

O'Neill, J. (ed.) (1973) *Modes of Individualism and Collectivism*, New York: St Martin's Press.

Orlikowski, W. J. (1992) 'The Duality of Technology: Rethinking the Concept of Technology in Organizations', *Organization Science*, 3(3): 398–427.

—— and Robey, D. (1991) 'Information Technology and the Structuring of Organizations', *Information Systems Research*, 2(2): 143–69.

Orwell, G. (1954) *Nineteen Eighty-Four*, Harmondsworth: Penguin.

O'Shea, D. (1993) 'MCI Shows Customers Proof of Savings', *Telephony*, 7 June.

Pahl, J. (1989) *Money and Marriage*, London: Macmillan.

Papanek, V. (1971) 'What is Design? A Definition of Design and the Function Complex', in V. Papanek (ed.), *Design for the Real World: Human Ecology and Social Change*, New York: Pantheon Books, 3–20.

Parnas, D. L., and Clements, P. C. (1986) 'A Rational Design Process: How and Why to Fake It', *IEEE Transactions on Software Engineering*, SE-12(2), 251–7.

Parry, J., and Bloch, M. (eds.) (1989) *Money and the Morality of Exchange*, Cambridge: Cambridge University Press.

Pavitt, K. (1984) 'Sectoral Patterns of Technical Change: Towards a Taxonomy and Theory', *Research Policy*, 13(6): 343–73.

Penrose, E. (1951) *The Economics of the International Patent System*, Baltimore: Johns Hopkins Press.

Peppers, D., and Rogers, M. (1993) *The One-to-One Future: Building Relationships One Customer at a Time*, New York: Doubleday.

Perez, C. (1983) 'Structural Change and the Assimilation of New Technologies in the Economic and Social System', *Futures* 15(4): 357–75.

Petronio, S. (1991) 'Communication Boundary Management: A Theoretical Model of Managing Disclosure of Private Information between Marital Couples', *Communication Theory*, 1(4): 311–35.

Petroski, H. (1994) *Design Paradigms: Case Histories of Error and Judgment in Engineering*. Cambridge: Cambridge University Press.

Pevsner, N. (1936) *Pioneers of Modern Design*, New York: Penguin.

Pinch, T., and Bijker, W. E. (1984) 'The Social Construction of Facts and Artefacts: or How the Sociology of Science and the Sociology of Technology Might Benefit Each Other', *Social Studies of Science*, 14: 399–441.

Piore, M. J., and Sabel, C. F. (1984) *The Second Industrial Divide: Possibilities for Prosperity*, New York: Basic Books.

Porat, M. U., and Rubin, M. R. (1977) *The Information Economy*, 9 vols., Washington, DC: Dept. of Commerce, Government Printing Office.

Porter, M. E., and Millar, V. E. (1985) 'How Information Gives You Competitive Advantage', *Harvard Business Review*, July–Aug.: 149–60.

Poster, M. (1990) *The Mode of Information: Poststructuralisms and Social Context*, Chicago: University of Chicago Press.

Powell, W. W. (1990) 'Neither Market nor Hierarchy: Network Forms of Organization', in B. M. Staw and L. L. Cummings (eds.), *Research in Organizational Behavior. An Annual Series of Analytical Essays and Critical Reviews* 12, Greenwich, Conn., JAI Press, 295–336.

Prahalad, C. K. (1993) 'The Role of Core Competencies in the Corporation', *Research Technology Management*, Nov./Dec., pp. 40–7.

Quintas, P. (1991*a*) 'Engineering Solutions to Software Problems: Some Institutional and Social Factors Affecting Change', *Technology Analysis and Strategic Management*, 3(4): 359–76.

—— (1991*b*) *Software Engineering: The International Policy Challenge*, Paris: OECD.

—— (ed.) (1993) *Social Dimensions of Systems Engineering: People, Processes, Policies and Software Development*, Hemel Hempstead: Ellis Horwood.

—— (1994*a*) ' "Ivory Tower Power": Multimedia and Society Feature', *Times Higher Education Supplement*, 4 Feb.

—— (1994*b*) 'A Product-Process Model of Innovation in Software Development', *Journal of Information Technology*, 9(1): 3–17.

—— (1994*c*) 'Programme Innovation? Trajectories of Change in Software Development', *Information Technology and People*, 7(1): 25–47.

—— (1994*d*) 'Software Engineering Policy and Practice: Lessons from the Alvey Programme', *Journal of Systems and Software*, 24(1): 67–88.

Rabel, G. (1948) 'Mathematical Instruments and Calculating Machines', *Science News VII*, Harmondsworth: Penguin, 112–24.

Raghavan, C. (1990) *Recolonization: GATT, the Uruguay Round & the Third World*. London/Penang: Zed Books Ltd/Third World Network.

Reddy, M. N. (1987) 'Technology, Standards, and Markets: A Market Institutionalization Perspective', in H. L. Gabel (ed.), *Product Standardization and Competitive Strategy*, Amsterdam: Elsevier, 47–66.

—— (1990) 'Product Self-Regulation: A Paradox of Technology Policy', *Technological Forecasting and Social Change*, 38: 49–63.

—— Cort, S. G., and Lambert, D. R. (1989) 'Industrywide Technical Product Standards', *R & D Management*, 19(1): 13–25.

Reich, R. B. (1992) *The Work of Nations: Preparing Ourselves for 21st Century Capitalism*, 2nd edn, New York: Vintage.

Reinbothe, J., and Howard, A. (1991) 'The State of Play in the Negotiations on TRIPS (GATT/Uruguay Round)', *European Intellectual Property Review*, 5: 157–64.

Rheingold, H. (1993) *Virtual Community: Homesteading on the Electronic Frontier*. Reading, Mass.: Addison-Wesley.

Rhodes, R. A. W. (1987) *The National World of Local Government*, London: Macmillan.

Robey, D., and Farrow, D. (1982) 'User Involvement in Information System Development: A Conflict Model and Empirical Test', *Management Science*, 28(1): 73–85.

Robins, K., and Webster, F. (1988) 'Cybernetic Capitalism: Information, Technology, Everyday Life', in V. Mosco and J. Wasko (eds.), *The Political Economy of Information*, Madison: University of Wisconsin Press, 44–75.

Rogge, J.-U., and Jensen, K. (1988) 'Everyday Life and Television and West Germany: An Empathetic-Interpretive Perspective on the Family as System', in J. Lull (ed.), *World Families Watch Television*, London: Sage, 80–115.

Ronfeldt, D. (1993) 'Institutions, Markets and Networks: A Framework About the Evolution of Societies', RAND Program for Research on Immigration Policy, Domestic Research Division, Report No. DRU-590-FF prepared for the Ford Foundation, Dec.

Roobeek, A. J. M. (1987) 'The Crisis in Fordism and the Rise of a New Technological Paradigm', *Futures*, 19(2): 129–54.

Rosenau, J. N. (1986) 'Before Cooperation: Hegemons, Regimes, and Habit-Driven Actors in World Politics', *International Organization*, 40(4): 849–94.

—— (1988) 'Patterned Chaos in Global Life: Structure and Process in the Two Worlds of Global Politics', *International Political Science Review*, 9(4): 327–64.

—— (1992) 'Governance, Order and Change in World Politics', in J. N. Rosenau and E. O. Czempiel (eds.), *Governance without Government: Order and Change in World Politics*, New York: Cambridge University Press, 1–29.

Rosenberg, E. S. (1976) 'Standards and Industry Self-Regulation', *California Management Review*, 19(1): 79–90.

Rosenberg, N. (1982) *Inside the Black Box: Technology and Economics*, Cambridge: Cambridge University Press.

—— (1994*a*) 'Critical Issues in Science Policy Research', in *Exploring the Black Box: Technology, Economics and History*, Cambridge: Cambridge University Press, 139–58, first published in 1991, *Science and Public Policy*, 18(6): 335–46.

—— (1994*b*) *Exploring the Black Box: Technology, Economics and History*, Cambridge: Cambridge University Press.

Rothwell, R. (1992) 'Successful Industrial Innovation: Critical Factors in the 1990s', *R & D Management*, 22(3): 221–39.

Ruggie, J. G. (1975) 'International Responses to Technology', *International Organization*, 29(3), 569–70.

—— (1992) 'Multilateralism: The Anatomy of an Institution', *International Organization*, 46(3), 561–98.

—— 'Territoriality and Beyond: Problematizing Modernity in International Relations', *International Organization*, 47(1), 139–74.

Ruggles, M. (1993) 'Mixed Signals: Personal Data Control in the Intelligent Network', *Media Information Australia*, 67 (Feb.): 28–9.

—— (1993) 'Territoriality and Beyond: Problematizing Modernity in International Relations', *International Organization*, 47(1): 139–74.

Ruthen, R. (1993) 'Adapting to Complexity', *Scientific American*, 268(1): 110–17.

Rutherford, M. (1989) 'What is Wrong with the New Institutional Economics (and What Is still Wrong with the Old)?', *Review of Political Economy*, 1(3): 299–318.

—— (1994) *Institutions in Economics: The Old and the New Institutionalism*, Cambridge: Cambridge University Press.

Sahal, D. (1981) 'Alternative Conceptions of Technology', *Research Policy*, 10(4): 368–402.

—— (1985) 'Technological Guideposts and Innovation Avenues', *Research Policy*, 14(2): 61–82.

Salter, L. (1985) 'Science and Peer Review: The Canadian Standard-Setting Experience', *Science, Technology and Human Values*, 10(4): 37–46.

—— (1988) *Mandated Science: Science and Scientists in the Making of Standards*, Dordrecht: Kluwer Academic Publishers.

—— and Hawkins, R. (1990) 'Problem Areas in the Management of Technology: Standards and Standards-Writing in Canada', in D. Wolfe and L. Salter (eds.), *Managing Technology—A Social Science Perspective*, Toronto: Garamond Academic Publishers, 137–63.

Samarajiva, R. (1994) 'Privacy in Electronic Public Space: Emerging Issues', *Canadian Journal of Communication*, 19(1): 87–99.

—— (1996) 'Consumer Protection in the Decentralized Network: A Mapping of the Research and Policy Terrain', in E. Noam (ed.), *Private Networks and Public Objectives*, Amsterdam: Elsevier, forthcoming.

—— and Shields, P. (1992) 'Emergent Institutions of the "Intelligent Network": Toward a Theoretical Understanding', *Media, Culture and Society*, 14(3): 397–419.

—— —— (1993) 'Institutional and Strategic Analysis in Electronic Space: A Preliminary Mapping', paper presented at the 43rd Conference of the International Communication Association, Washington, DC, 27–31 May.

Samuels, W. J. (1990) 'Comments: The Old Versus the New Institutionalism', *Review of Political Economy*, 2(1): 83–6.

Samuelson, P. A. (1954) 'The Pure Theory of Public Expenditure', *Review of Economics and Statistics*, 36: 387–89.

Savage, J. G. (1989) *The Politics of International Telecommunications Regulation*, Boulder, Colo.: Westview Press.

Saviotti, P. P., and Metcalfe, J. S. (1984) 'A Theoretical Approach to the Construction of Technological Output Indicators', *Research Policy*, 13(3): 141–51.

—— (1993) 'The Theoretical Underpinnings of the Concept of National System of Innovation (NSI)', paper presented at the EUNETIC Workshop on Globalization versus National or Local Systems of Innovation, Strasbourg, 11–12 Mar.

Sayers, A. (1992) *Method in Social Science: A Realist Approach*, 2nd edn., New York: Routledge.

Scannell, P. (1989) 'Public Service Broadcasting and Modern Public Life', *Media, Culture and Society*, 11(2): 135–66.

—— and Cardiff, D. (1991) *A Social History of British Broadcasting i: 1922–1939*, Oxford: Blackwell.

Scharpf, F. W. (1993) 'Coordination in Hierarchies and Networks', in F. W. Scharpf (ed.), *Games in Hierarchies and Networks: Analytical and Empirical Approaches to the Study of Governance Institutions*, Frankfurt am Main: Campus Verlag, 125–65.

Schmidt, S. K., and Werle, R. (1992) 'The Development of Compatibility Standards in Telecommunications: Conceptual Framework and Theoretical Perspective', in M. Dierkes and U. Hoffmann (eds.), *New Technology at the Outset: Social Forces in the Shaping of Innovations*, Frankfurt am Main: Campus Verlag, 301–31.

Schwach, V. (1992) 'L'Intégration des Objets Techniques dans la Vie Quotidienne', *Sociologie des Techniques de la Vie Quotidienne*, Paris: Éditions l'Harmattan, Collections Logiques Sociales, 103–8.

Science Policy Research Unit (1995) 'Centre for Information and Communication Technologies: Bibliography 1986–1995', Brighton: University of Sussex.

Scott-Morton, M. (ed.) (1991) *The Corporation of the 1990s: Information Technology and Organizational Transformation*, Oxford: Oxford University Press.

Seidman, H. (1980) *Politics, Position, and Power: The Dynamics of Federal Organization*, New York: Oxford University Press.

Senker, J. (1996) 'Networks and Tacit Knowledge in Innovation', *Economies et Sociétés*, forthcoming.

Sharrock, W., and Anderson, B. (1991) 'The User as a Scenic Feature of the Design Space', Technical Report EPC-91-105, Rank Xerox EuroPARC, Cambridge.

—— and Button, G. (1991) 'The Social Actor: Social Action in Real Time', in G. Button (ed.), *Ethnomethology and the Human Sciences*, Cambridge: Cambridge University Press, 137–75.

Shields, P. (1996) 'Beyond Individualism and the Ecology of Games: Structures, Institutions and Communication Policy', *Communication Theory*, forthcoming.

—— Dervin, B., Richter, C., and Soller, R. E. (1993) 'Who Needs "POTS-Plus" Services? A Comparison of Residential User Needs Along the Rural-Urban Continuum', *Telecommunications Policy*, 17(8): 563–88.

Silverstone, R. (1994a) *Television and Everyday Life*, London: Routledge.

—— (1994*b*) 'Time, Information and Communication Technologies and the Household', *Time and Society*, 3(2): 283–311.

—— and Haddon, L. (1993) 'Future Compatible? Information and Communication Technologies in the Home: A Methodology and a Case Study', Report Prepared for the Commission of the European Communities Socio-Economic and Technical Impact Assessments and Forecasts, RACE Project 2086, Brighton: Science Policy Research Unit, University of Sussex.

—— and Hirsch, E. (1992*a*) 'Introduction', in R. Silverstone and E. Hirsch (eds.), *Consuming Technologies: Media and Information in Domestic Spaces*, London: Routledge, 1–14.

—— —— (eds.) (1992*b*) *Consuming Technologies: Media and Information in Domestic Spaces*, London: Routledge.

—— and Morley, R. (1990) 'Families and their Technologies: Two Ethnographic Portraits', in T. Putnam and C. Newton (eds.), *Household Choices*, London: Futures Publications, 74–83.

—— Hirsch, E., and Morley, D. (1992) 'Information and Communication Technologies and the Moral Economy of the Household', in R. Silverstone and E. Hirsch (eds.), *Consuming Technologies: Media and Information in Domestic Spaces*, London: Routledge 15–31.

Simmel, G. (1904) 'Fashion', *International Quarterly*, 10: 130–55.

Simon, H. A. (1945) *Administrative Behavior: A Study of Decision-Making Processes in Administrative Organization*, New York: Free Press.

—— (1955) 'A Behavioral Model of Rational Choice', *Quarterly Journal of Economics*, 69: 99–118.

—— (1988) 'Human Nature in Politics: The Dialogue of Psychology with Political Science', in M. Capanella (ed.), *Between Rationality and Cognition: Policy-Making Under Conditions of Uncertainty, Complexity and Turbulence*, Torino: Albert Menier, 11–35.

Simpson, J. A., and Weiner, E. S. C. (eds.) (1989) *The Oxford English Dictionary*, iv, 2nd edn., Oxford: Clarendon Press.

Sinclair, B. (1969) 'At the Turn of a Screw: William Sellers, The Franklin Institute, and a Standard American Thread', *Technology and Culture*, 10(1): 20–34.

Skinner, D. (1992) 'Technology, Consumption and the Future: The Experience of Home Computing', unpublished PhD Diss., Brunel University, Uxbridge.

Slater, J. (1967) *The OAS and United States Foreign Policy*, Columbus, Oh.: Ohio State University Press.

Smith, D. (1978) ' "K is Mentally Ill": The Anatomy of a Factual Account', *Sociology*, 12(1): 23–53.

Smythe, D. W. (1972) 'The "Orbital Parking Slot" Syndrome and Radio Frequency Management', *Quarterly Review of Economics and Business*, Summer, 7–17.

—— (1981) *Dependency Road: Communications, Capitalism, Consciousness, and Canada*, Norwood, NJ: Ablex.

—— (1987) 'Radio Spectrum Policy and World Needs', *Prometheus*, 5(2): 263–83.

—— (1989) 'Notes on a Search for the Reality of Information and Communication', unpub. MS, Simon Fraser University, Burnaby, BC, Canada, 10 Nov.

—— (1991) 'Theory about Communication and Information', unpub. MS, Simon Fraser University, Burnaby, BC, Canada.

Soja, E. W. (1989) *Postmodern Geographies: The Reassertion of Space in Critical Social Theory*, London: Verso.

Soja, E. W. (1980) 'The Socio-spatial Dialectic', *Annals of the Association of American Geographers*, 70: 175–90.

Sommerville, I. (1992) *Software Engineering*, 4th edn., Wokingham: Addison-Wesley.

Soroos, M. S. (1982) 'The Commons in the Sky: The Radio Spectrum and Geosynchronous Orbit as Issues in Global Policy', *International Organization*, 36(3): 665–85.

Stansky, P. (1985) *Redesigning the World: William Morris, the 1880s, and the Arts and Crafts*: Princeton, NJ: Princeton University Press.

Stehr, N., and Ericson, R. V. (1992*a*) 'The Culture and Power of Knowledge in Modern Society', in N. Stehr and R. V. Ericson (eds.), *The Culture and Power of Knowledge: Inquiries into Contemporary Societies*, Berlin/New York: Walter de Gruyter, 3–19.

—— —— (eds.) (1992*b*) *The Culture and Power of Knowledge: Inquiries into Contemporary Societies*, Berlin/New York: Walter de Gruyter.

Stein, D. L. (ed.) (1989) *Lectures in the Sciences of Complexity*, New York: Addison-Wesley.

Sterling, B. (1992) *The Hacker Crackdown*, London: Penguin.

Stigler, G. (1961) 'The Economics of Information', *Journal of Political Economy*, 69(3): 213–26.

Stones, R. (1991) 'Strategic Context Analysis: A New Research Strategy for Structuration Theory', *Sociology*, 25(4): 673–95.

Strassmann, P. A. (1994) 'The Politics of Electronic Commerce', paper presented at the 5th World Congress of EDI Users, Brighton, 16 June.

Strauss, A. L. (1982) 'Social Worlds and Legitimation Processes', in N. Denzin (ed.), *Studies in Symbolic Interaction*, iv, Greenwich, Conn.: JAI Press, 171–90.

—— (1984) 'Social Worlds and Their Segmentation Processes', in N. Denzin (ed.), *Studies in Symbolic Interaction*, v, Greenwich, Conn.: JAI Press, 123–39.

Sung, L. (1992) 'WARC-92: Setting the Agenda for the Future', *Telecommunications Policy*, 16(8): 624–34.

Susman, G. I. (1992) 'Integrative Design and Manufacturing for Competitive Advantage', in G. I. Susman (ed.), *Integrative Design and Manufacturing for Competitive Advantage*, New York/Oxford: Oxford University Press, 3–11.

Swann, P. (1994) 'Reaching Compromise in Standards Setting Institutions', in G. Pogorel (ed.), *Global Telecommunications Strategies and Technological Changes*, Amsterdam: North-Holland, 241–53.

SWTRHA (1993) *Report of the Inquiry in the London Ambulance Service*, South West Regional Health Authority, London.

Talbot, D. E., and Witty, R. W. (1983) *Alvey Programme Software Engineering Strategy*, London: Alvey Directorate, Dept. of Trade and Industry.

Tang, P. (1995) 'Multimedia Services: A Need for "Superhighway Cops"?', paper prepared for The 'Virtual Future': Policies for Multimedia in the 21st Century Colloquium, Science Policy Research Unit, University of Sussex, London, 20 Mar.

Tassey, G. (1991) 'The Functions of Technology Infrastructure in a Competitive Economy', *Research Policy*, 20(4): 345–61.

Taylor, J. R., and E. J. Van Every (1993) *The Vulnerable Fortress: Bureaucratic Organisation and Management in the Information Age*, Toronto: University of Toronto Press.

Taylor, P. (1990) 'Regionalism and Functionalism Reconsidered: A Critical Theory', in A. J. R. Groom and P. Taylor (eds.), *Frameworks for International Co-operation*, London: Pinter, 234–53.

Teece, D. J. (1986) 'Profiting from Technological Innovation: Implications for Integration, Collaboration, Licensing and Public Policy', *Research Policy*, 15(6): 285–305.

—— (1988) 'Technological Change and the Nature of the Firm', in G. Dosi, C. Freeman, R. Nelson, G. Silverberg, and L. Soete (eds.), *Technical Change and Economic Theory*, London: Pinter, 256–81.

Telecommunications Council (1994) 'Reforms Toward the Intellectually Creative Society of the 21st Century', Ministry of Posts and Telecommunications, Tokyo, 31 May.

Tennant, F. R. (1956) *Philosophical Theology*, Cambridge: Cambridge University Press.

Thomas, G., and Miles, I. (1990) *Telematics in Transition: The Development of New Interactive Services in the United Kingdom*, Harlow: Longman.

Thompson, E. P. (1955) *William Morris: Romantic to Revolutionary*, London, Merlin Press.

—— (1971) 'The Moral Economy of the English Crowd in the Eighteenth Century', *Past and Present*, 50: 76–136.

—— (1978) *The Poverty of Theory and Other Essays*, New York: Monthly Review Press.

Thompson, G. V. (1954) 'Intercompany Technical Standardization in the Early American Automobile Industry', *Journal of Economic History*, 14(1): 1–20.

Thrall, C. A. (1982) 'The Conservative Use of Modern Household Technology', *Technology and Culture*, 23(2): 175–94.

Traxler, F., and Unger, B. (1994) 'Governance, Economic Restructuring, and International Competitiveness', *Journal of Economic Issues*, 28(1): 1–23.

Trebing, H. M. (1994) 'The Networks as Infrastructure: The Re-establishment of Market Power', *Journal of Economic Issues*, 28(2): 379–89.

Turkle, S. (1986) *The Second Self: Computers and the Human Spirit*, London: Granada.

—— (1988) 'Computational Reticence: Why Women Fear the Intimate Machine', in C. Kramarae (ed.), *Technology and Women's Voices: Keeping in Touch*, London: Routledge & Kegan Paul, 41–61.

UNCTAD (1991) 'Trade and Development Report', UNCTAD/IRD/II, Sales No. E. 91.II.D.15, Geneva: United Nations Conference on Trade and Development.

United States (1993a) 'National Information Infrastructure: Agenda for Action: Realizing the Information Future', Information Infrastructure Task Force, Washington, DC, 15 Sept.

—— (1993b) 'President Clinton, Vice-President Gore, Technology for America's

Economic Growth: A New Direction to Build Economic Strength', Office of the President, Washington, DC, 22 Feb.

United States (1994) 'US Industrial Outlook', Dept. of Commerce, Washington, DC.

United States Congress (1994) 'Digital Telephony Bill', House of Representatives HR 4922, 103rd Session.

United States International Trade Commission (1991) 'Global Competitiveness of US Advanced-Technology Manufacturing Industries: Communications Technology and Equipment', United States International Trade Commission Publication 2439, Washington, DC.

United States Senate (1990) 'United States Senate Report No. 101-456 101st Cong., 2nd Sess, on S.2800,' Communications, Competitiveness and Infrastructure Modernization Act of 1990, at n.3, 20-21 quoting Senator Burns, cited in H. Geller (1991) 'Fibre Optics: An Opportunity for a New Policy', Report of the Annenberg Washington Program, Communication Policy Studies, Northwestern University, Chicago.

Veblen, T. (1898) 'Why is Economics Not an Evolutionary Science?', in T. Veblen, *The Place of Science in Modern Civilization*, New York: Russell & Russell (repub. 1961, 56–81).

—— (1899) *The Theory of the Leisure Class: An Economic Study of Institutions*, New York: Mentor (repub. 1953).

—— (1909) 'The Limitations of Marginal Utility', in T. Veblen (ed.), *The Place of Science in Modern Civilization*, New York: Russell & Russell (repub. 1961), 231–51.

—— (1919) *The Vested Interests and the Common Man*, New York: Augustus M. Kelley (repub. 1964).

Verman, L. C. (1973) *Standardization: A New Discipline*, Hamden, Conn.: Archon Books.

Vernon, R. (1966) 'International Investment and International Trade in the Product Cycle', *Quarterly Journal of Economics*, 80(2): 190–207.

Viswanath, K., Samarajiva, R., and Park, E. (1994) 'Covering Telecommunication Privacy: A Study of National and Local Press Coverage of the Policy Process', paper presented at the 19th Conference of the International Association for Mass Communication Research, Seoul, Korea, July.

von Hippel, E. (1977) 'Transferring Process Equipment Innovations from User-Innovators to Equipment Manufacturing Firms', *R&D Management*, 8(1): 13–22.

von Tunzelmann, N. (1994) 'Interactions between Technological and Economic Factors: A Study of Three Industries', paper presented at the EUNETIC Conference, Evolutionary Economics of Technical Change: Assessment of results and new frontiers European Parliament, Strasbourg, 6–8 Oct.

Waldrop, M. M. (1992) Complexity: *The Emerging Science at the Edge of Order and Chaos*, New York: Simon & Schuster.

Wallenstein, G. (1990) *Setting Global Telecommunications Standards: The Stakes, the Players and the Process*, Dedham: Artech House.

Weber, M (1978) *Selections in Translations*, ed. W. G. Runciman, Cambridge: Cambridge University Press.

Webster, J., and Williams, R. (1993) 'Mismatch and Tension: Standard Packages and Non-Standard Users', in P. Quintas (ed.), *Social Dimensions of Systems Engineering: People, Processes, Policies and Software Development*, Hemel Hempstead: Ellis Horwood, 179–96.

Weidlein, E. R., and Reck, V. (1956) 'A Million Years of Standards', in D. Reck (ed.), *National Standards in a Modern Economy*, New York: Harper & Bros., 5–20.

Weisman, L. K. (1992) *Discrimination by Design: A Feminist Critique of the Manmade Environment*, Urbana, Ill.: University of Illinois Press.

Weiss, M. B. H., and Sirbu, M. (1990) 'Technological Choice in Voluntary Standards Committees: An Empirical Analysis', *Economics of Innovation and New Technology*, 1(1): 111–33.

Weiss, M. J. (1988) *The Clustering of America*, New York: Harper & Row.

Wellenius, B. (1993) *Telecommunications: World Bank Experience and Strategy*, Washington, DC: World Bank.

Wendt, A. E. (1987) 'The Agent-Structure Problem in International Relations Theory', *International Organisation*, 41(3): 335–70.

—— (1992) 'Anarchy is What States Make of It: The Social Construction of Power Politics', *International Organization*, 46(2): 391–425.

Whitworth, J. (1882) *Papers on Mechanical Subjects: Part 1 True Planes, Screw Threads, and Standard Measures*, London: E. F. & N. Spon.

Wiener, N. (1950) *The Human Use of Human Beings: A Cybernetic Approach*, New York: Houghton Mifflin.

Wilkinson, B. (1983) *The Shopfloor Politics of New Technology*, London: Heinemann.

Williams, R. (1958) *Culture & Society: Coleridge to Orwell*, London: Hogarth Press.

—— (1974) *Television: Technology and Cultural Form*, London: Fontana.

—— (1980) 'Advertising: The Magic System', in R. Williams, *Problems in Materialism and Culture: Selected Essays*, London: Verso, 170–95.

—— (1983) *Towards 2000*, London: Hogarth Press.

Winograd, T. (1991) 'Debate on a paper by J. S. Brown, "Research that Reinvents the Corporation" ', *Harvard Business Review*, Jan.–Feb., 172.

Winter, S. G. (1987) 'Knowledge and Competence as Strategic Assets', in D. J. Teece (ed.), *The Competitive Challenge: Strategies for Industrial Innovation and Renewal*, Cambridge, Mass.: Ballinger, 159–84.

Wired (1994) 'All Music', *Wired*, Feb., 52–56.

Womack, J. P., Jones, D. T., and Roos, D. (1990) *The Machine that Changed the World*, New York: Rawson Associates.

Woolgar, S. (1991) 'Configuring the User: The Case of Usability Trials', in J. Law (ed.), *A Sociology of Monsters*, London: Routledge, 57–99.

—— (1993) 'Software is Society made Malleable: The Importance of Conceptions of Audience in Software in Research Practice', PICT Policy Research Paper No. 25, Brunel University, Nov., Uxbridge.

World Intellectual Property Organization (1983) 'Committee of Experts on the Legal Protection of Computer Software', Report Adopted by the Committee of Experts, Second Session, LPCS/II/6, Geneva: World Intellectual Property Organization, 13–17 June.

—— (1990) 'Principles for a Draft Treaty for the Settlement of Disputes between States in the Field of Intellectual Property', Memorandum prepared by the

International Bureau SD/CE/II/2, Geneva: World International Property Organization, 14 Aug.

World Intellectual Property Organization (1992) 'Report on the Activities of WIPO in the Year 1992', Geneva: World Intellectual Property Organization.

Wright, M. (1988) 'Policy Community, Policy Network and Comparative Industrial Policies', *Political Studies*, 36: 593–612.

Yusuf, A. A., and von Hase, A. M. (1992) 'Intellectual Property Protection and International Trade: Exhaustion of Rights Revisited', *World Competition*, 16(1): 115–31.

Zartman, I. W. (1977) 'Negotiation as a Joint Decision-making Process', in I. W. Zartman (ed.), *The Negotiation Process: Theories and Applications*, Beverly Hills, Calif.: Sage, 67–86.

Zysman, J. (1994) 'How Institutions Create Historically Rooted Trajectories of Growth', *Industrial and Corporate Change*, 3(1): 243–83.

INDEX